I0058161

Combinatorics Problems and Solutions
by Stefan Hollos and J. Richard Hollos
ISBN 978-1-887187-13-8

Abrazol Publishing

an imprint of Exstrom Laboratories LLC
662 Nelson Park Drive, Longmont, CO 80503-7674 U.S.A.

About the Cover

Cover created with the help of POV-Ray and Inkscape. We thank the creators and maintainers of this excellent software.

Contents

Preface   1

1 Introduction   5

    1.1   Definitions   5

    1.2   A Guide to Solving Counting Problems .   8

    1.3   Sum and Product Rule   16

    1.4   Permutations and Combinations   17

    1.5   Inclusion Exclusion Principle   21

    1.6   Stirling Numbers of the Second Kind . .   25

2 Problems   29

3 Exercises   65

Further Reading   127

Acknowledgements   129

About the Authors   131

Why would anyone want to solve combinatorics problems? The best reason has to be because it's simply fun. It's also a great way to sharpen your problem solving skills and give your brain a good workout. An hour spent solving combinatorics problems is better than an hour spent playing chess since it's not only fun but it gives you skills you can use in mathematics, computer science, physics, biology, etc.

Let's define what we mean by combinatorics. Much of combinatorics is based on the idea of counting. This may involve simply counting the number of elements in a set. It sounds simple but it's not a matter of just looking and counting. The set is usually only defined as elements that meet some condition or have some property. You could, in principle, construct the set from the definition and then count the number of elements it contains. When the number of elements runs into the thousands, millions, billions, or becomes infinite then things become quite tedious. So to solve combinatorics problems you often need some insight, ingenuity, and the ability to turn the problem into a different form that is more easily solved. What could be more fun than that?

This is not meant to be a textbook on combinatorics but there is enough introductory material so that even someone with little or no prior exposure to the subject can get something out of it. Some familiarity with the concept of sets, subsets, factorials, and basic algebra is all that is required. We start with some definitions in the introduction along with a guide to solving counting problems. The guide is a list of some of the most common counting problems. There is an equation for each problem and a set of equivalent descriptions of the problem. This is followed by short sections that explain some of the most basic principles in combinatorics. They were chosen because they are used in the problems but they are by no means exhaustive.

After the introduction comes the problems and exercises. The problems are generally easier and shorter than the exercises and increase in complexity as you go. Some of the exercises can be quite involved and may take some time to fully work out. A computer algebra system capable of dealing with very large numbers may be helpful for some of the problems and exercises (Emacs calc also works). Each problem and exercise is fully worked out in detail. Most of the problems and exercises are modernized versions of the problems and exercises found in the book: *Choice and Chance* by W. A. Whitworth (see Further Reading at the end of the book). Happy problem solving.

We can be reached by email at:
stefan[at]exstrom DOT com
richard[at]exstrom DOT com

Stefan Hollos and J. Richard Hollos
Exstrom.com
QuantWolf.com
Exstrom Laboratories LLC
Longmont, Colorado, U.S.A.
January 2013

1.1 Definitions

This is a short list of definitions so that you know what we're talking about. You probably know this stuff already but it's a good idea to skim over it to avoid any possible confusion and to make sure we're both talking the same language.

- A set is a collection of objects called elements or members. All the elements of a set are unique. The elements of a set are not ordered. The set of the first four letters of the English alphabet can be written as $\{a, b, c, d\}$, or $\{d, c, b, a\}$, or any other order. The number of elements in a set is called its size or cardinality.

- The set B is a subset of the set A if every element of B is also an element of A. The empty set (a set with no elements) is a subset of every set. A subset can be constructed as an unordered sampling of elements from a set.

- The set $[n] = \{1, 2, 3, \ldots, n\}$ is a subset of the set of integers.

- A multiset is a collection of objects where each object may occur one or more times. $\{a, a, a, b, c, c, d\}$

is a multiset with 3 copies of element a, 1 copy of b, 2 copies of c, and 1 of d. Like a set, a multiset is not ordered. A multiset can be constructed as an unordered sampling with replacement from a set.

- A list is an ordered collection of elements. The list $[a, b, c, d]$, differs from the list $[d, c, b, a]$. A list may also be called a permutation of the elements it contains. A list can be constructed as an ordered sampling of elements from a set.

- A circular list is an ordered collection of elements. The order must be unique with respect to a circular shift of the elements. The lists $[a, b, c]$, $[b, c, a]$, and $[c, a, b]$ are all the same circular list. A circular list may also be called a circular permutation of the elements it contains.

- An alphabet is a set of symbols also called letters.

- A multi-alphabet is a multiset of symbols, i.e. it may contain one or more copies of each symbol.

- A word is a list of symbols usually drawn from a specified alphabet or multi-alphabet. The number of symbols in a word is called its length or size.

- A partition of the set A into k parts is a set of k subsets of A with each element of A appearing in exactly one of the subsets and none of the subsets equal to the empty set.

- A composition of the integer n into k parts is a list of integers of size k that sum to n. Some of the elements (integers) may equal zero.

- A weak composition is a composition where none of the elements is zero.

- A partition of the integer n into k parts is a set of integers of size k that sum to n. None of the elements (integers) may equal zero.

- The partition number $p_k(n)$ is the number of ways to partition n into k parts.

- The Stirling number of the second kind $S(n, m)$ is the number of ways to partition a set of size n into m nonempty subsets. It is defined as follows:

$$S(n, m) = \frac{1}{m!} \sum_{k=0}^{m} (-1)^k \binom{m}{k} (m - k)^n$$

where $\sum_{k=0}^{m}$ means to sum what follows from $k = 0$ to m. The remaining notation in this expression is explained in the section Permutations and Combinations.

1.2 A Guide to Solving Counting Problems

The following equations are for counting problems. Each equation is followed by one or more questions that it answers. The twelve equations: 1.1, 1.2, 1.4, 1.5, 1.6, 1.8, 1.9, 1.10, 1.11, 1.12, 1.13, 1.14, are for counting problems known as the twelvefold way. This is an attempt to create a systematic classification of counting problems (see Richard P. Stanley, Enumerative Combinatorics, Vol 1, 2nd edition, p. 79). One intuitive way of framing these problems is in terms of placing balls into bins (or urns). Both balls and bins can be distinct or indistinct (identical) and there are three ways the balls can be placed (mapped) into bins. The four distinct, indistinct combinations and the three mappings produce the twelvefold way. Additional problems that do not fit neatly into the twelvefold way are included, making it the fifteenfold way.

Probably the best way to learn the formulas and their applicability in the fifteenfold way below is not to intentionally try to memorize them, but to learn them in the process of working problems. When starting on a new problem, think of the essential model that the problem represents, then find the matching question, and that will give you the formula. As you work more problems, you'll remember more.

$$m^n \tag{1.1}$$

- How many ways can you place n distinct balls into m distinct bins with no restrictions?

- How many n letter words can you make with m types of letters?

- How many functions are there from $[n]$ to $[m]$.

$$\frac{m!}{(m-n)!} \tag{1.2}$$

- How many ways can you place n distinct balls into m distinct bins with no more than one ball in each bin?

- How many lists of size n can you construct from a set of size m?

- How many permutations are there of m distinct things taken n at a time?

- How many words of length n can you construct from an alphabet of size m?

$$\frac{m!}{n(m-n)!} \tag{1.3}$$

- How many circular lists of size n can you construct from a set of size m?

- How many different bracelets can you make using n beads from a set of m beads?

$$m!S(n,m) \tag{1.4}$$

- How many ways can you place n distinct balls into m distinct bins with at least one ball in each bin?

- How many ways can you create m lists from a set of size n with each list being greater than or equal to one in length?

$$\binom{n + m - 1}{m - 1} \qquad (1.5)$$

- How many ways can you place n identical balls into m distinct bins with no restrictions?

- How many multisets of size n can you create by sampling with replacement from a set of size m?

- How many compositions of n into m parts are there?

$$\binom{m}{n} \qquad (1.6)$$

- How many ways can you place n identical balls into m distinct bins with no more than one ball in each bin?

- How many subsets of size n can you produce from a set of size m?

$$\sum_{n=0}^{m} \binom{m}{n} = 2^m \qquad (1.7)$$

- How many subsets can you create from a set of size m?

- How many words of length m can you make using only two kinds of letters?

- How many m digit binary numbers are there?

$$\binom{n-1}{m-1} \qquad (1.8)$$

- How many ways can you place n identical balls into m distinct bins with at least one ball in each bin?

- How many compositions of n into m nonzero parts are there?

$$\sum_{k=1}^{m} S(n,k) \qquad (1.9)$$

- How many ways can you place n distinct balls into m identical bins with no restrictions?

$$\begin{array}{ll} 1 & \text{if } n \leq m \\ 0 & \text{if } n > m \end{array} \qquad (1.10)$$

- How many ways can you place n distinct balls into m identical bins with no more than one ball in each bin?

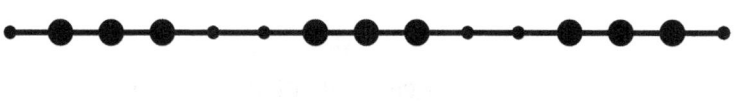

$$S(n,m) \qquad (1.11)$$

- How many ways can you place n distinct balls into m identical bins with at least one ball in each bin?

- How many ways can you partition a set of size n into m nonempty subsets?

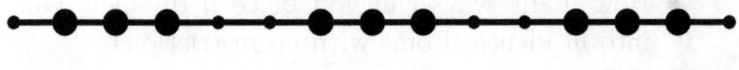

$$\sum_{k=1}^{m} p_k(n) \qquad (1.12)$$

- How many ways can you place n identical balls into m identical bins with no restrictions?

$$\begin{array}{ll} 1 & \text{if } n \le m \\ 0 & \text{if } n > m \end{array} \qquad (1.13)$$

- How many ways can you place n identical balls into m identical bins with no more than one ball in each bin?

$$p_m(n) \qquad\qquad (1.14)$$

- How many ways can you place n identical balls into m identical bins with at least one ball in each bin?

$$\frac{n!}{k_1!k_2!\cdots k_m!} \qquad\qquad (1.15)$$

- How many ways can you place n distinct balls into m distinct bins so that bin i contains k_i balls with $k_1 + k_2 + \cdots + k_m = n$?

- How many ways can you order a multiset of n objects of m different types with k_i objects of type i?

- How many words of length n can you construct using a multi-alphabet with m kinds of letters and k_i copies of letter i?

- How many ways can you split a set of n objects into m subsets with the sizes of the subsets given by k_i, $i = 1, 2, \ldots, m$?

1.3   Sum and Product Rule

The sum and product rules are two of the most basic principles of counting. The sum rule says that if you have two or more sets then the number of ways of selecting an element from one or the other of the sets is the sum of the sizes of the sets. For k sets with sizes n_1 through n_k the number of ways of selecting an element from any one of the sets is

$$N = n_1 + n_2 + \cdots + n_k \qquad (1.16)$$

This formula is based on the assumption that there is no overlap between the sets, i.e. the same element does not appear in more than one set. Another way of saying this is that the intersection of the sets must be the empty or null set. If there is a nonempty intersection then the inclusion exclusion principle (discussed below) can be used. It is also possible that the same element appears in more than one set and each appearance has to be counted in which case the formula is also valid.

The product rule says that if you have k sets with sizes n_1 through n_k then the number of ways of selecting one element from each of the sets is:

$$N = n_1 \cdot n_2 \cdot n_3 \cdots n_k \qquad (1.17)$$

The sum and product rules are often used together. Suppose there are three sets A, B, and C with sizes n_A, n_B, and n_C. How many ways can two elements be selected from two different sets? The product rule says that the number of ways of selecting one element from A and one from B is $n_A \cdot n_B$. This is one set of possibilities. You can also select one element from A and one from C in $n_A \cdot n_C$ ways. This is another set of possibilities. Finally you can make one selection from B and one from C in $n_B \cdot n_C$ ways. You have three sets of ways to make the selection and you have to pick from one of them so the sum rule says the total number of ways to make the selection is

$$N = n_A \cdot n_B + n_A \cdot n_C + n_B \cdot n_C \qquad (1.18)$$

1.4   Permutations and Combinations

The number of permutations of n objects is the number of ways they can be arranged in linear order. There are n choices for the first in line, $n-1$ choices for the second, and so on. The number of permutations is then $n! = n \cdot (n-1) \cdot (n-2) \cdots 2 \cdot 1$. The number of permutations of n objects taken k at a time is $\frac{n!}{(n-k)!}$. This is the number of ways that k objects can be selected from n objects where the order of selection matters. It is the number of lists of length k that can be made from a set of n objects.

In a circular permutation the elements are arranged in a circular order. There is no natural beginning or end of the arrangement and any circular shift of the elements around the circle does not count as a new arrangement. There are $(n-1)!$ circular permutations of n elements as opposed to $n!$ linear permutations. To see this, pick a point on the circle to correspond with the start of a linear permutation. If you put one of the linear permutations on the circle then every one of its n possible circular shifts will correspond to a different linear permutation when read from the starting point but each of these shifts is counted as the same circular permutation. For every circular permutation there are n linear permutations therefore the number of circular permutations must be $\frac{n!}{n} = (n-1)!$.

The same argument applies to the number of circular permutations of n elements taken k at a time. There are k circular shifts of the k elements that correspond to different linear permutations but the same circular permutation. So to get the number of circular permutations, divide the linear permutations by k to get $\frac{n!}{k(n-k)!}$.

A combination of n elements taken k at a time is a k element subset of the n elements. It differs from a permutation of n elements taken k at a time in that the order of the k elements is irrelevant. The number of such combinations is

$$\binom{n}{k} = \frac{n!}{k!(n-k)!} \qquad (1.19)$$

This is equal to the number of k element subsets that can be formed from a set of n elements (the order of elements in a set or subset is irrelevant).

The term $\binom{n}{k}$ is called a binomial coefficient since it appears in the expansion of binomials. If you multiply out the binomial $(x+1)^n$ then the coefficient of x^k in the expansion will be $\binom{n}{k}$. For example in $(x+1)^5 = x^5 + 5x^4 + 10x^3 + 10x^2 + 5x + 1$ the coefficient of x^3 is $\binom{5}{3} = 10$ which also happens to equal the coefficient of x^2. In general $\binom{n}{k} = \binom{n}{n-k}$. From their definition in terms of a binomial expansion it is clear that:

$$\sum_{k=0}^{n} \binom{n}{k} = 2^n \qquad (1.20)$$

A useful recurrence for calculating binomial coefficients is:

$$\binom{n}{k} = \binom{n-1}{k-1} + \binom{n-1}{k} \qquad (1.21)$$

Suppose an n element set is to be divided into 3 subsets with k elements in the first subset, l in the second, and $n - k - l$ in the third. Start by dividing the set into a k element subset and a $n - k$ element subset. This can be done in $\binom{n}{k}$ ways. Next select the l elements for the second subset from the $n - k$ element subset. This can be done in $\binom{n-k}{l}$ ways. Now we have 3 subsets of the original n elements of size k, l, and $n - k - l$. The number of ways of doing this is:

$$\binom{n}{k}\binom{n-k}{l} = \frac{n!}{k!(n-k)!}\frac{(n-k)!}{l!(n-k-l)!} = \frac{n!}{k!l!(n-k-l)!}$$
$$(1.22)$$

The extension to more than 3 subsets is clear. Lets say we want r subsets with k_i elements in subset $i = 1, 2, \ldots, r$. This should completely divide up the n elements so that $k_1 + k_2 + \cdots + k_r = n$. The number of ways the subsets can be formed is

$$\binom{n}{k_1}\binom{n-k_1}{k_2}\cdots\binom{n-k_1-k_2-\cdots-k_{r-1}}{k_r}$$
$$(1.23)$$

When this equation is written out and simplified it becomes

$$\frac{n!}{k_1!k_2!\cdots k_r!} = \binom{n}{k_1, k_2, \cdots, k_r} \qquad (1.24)$$

The term on the right side of this equation is called a multinomial coefficient. The coefficients appear when you expand a multinomial. In this case they would appear in the expansion of $(x_1 + x_2 + \cdots + x_r)^n$. The coefficient of $x_1^{k_1} x_2^{k_2} \cdots x_r^{k_r}$ in the expansion is given by equation 1.24.

1.5 Inclusion Exclusion Principle

The Inclusion Exclusion Principle is used to count the number of elements in a collection of sets when there is some overlap between the sets, i.e. some elements appear in more than one set. The simplest example has two sets as shown in figure 1.1. Set A_1 has 7 elements, A_2 has 5 elements and the two sets have elements, 8 and 4, in common. If we sum the number of elements in each set then the elements they have in common will be counted twice. So the number they have in common has to be subtracted. The number of elements in A_1 or A_2 is then $7 + 5 - 2 = 10$. To get a general formula for two sets, let $|A_i|$ be the number of elements in set A_i then the number of elements in either set A_1 or set A_2 is given by:

$$|A_1 \cup A_2| = |A_1| + |A_2| - |A_1 \cap A_2| \qquad (1.25)$$

where $|A_1 \cap A_2|$ is the number of elements in both sets A_1 and A_2 i.e. the number of elements in the intersection of the two sets.

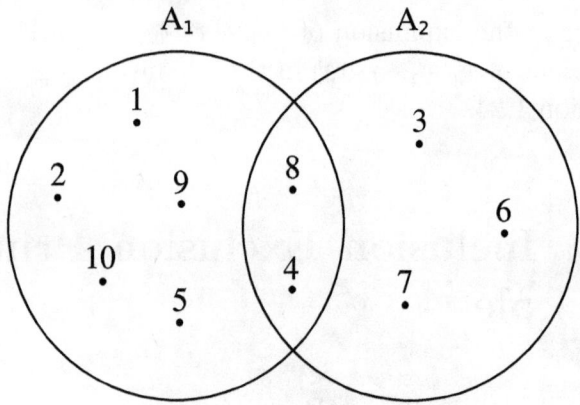

Figure 1.1: Two set example.

To extend this to three sets we need not just the number of elements in two sets at once but also the number of elements in three sets at once. Figure 1.2 shows an example. It is similar to the two set example but now there is a third set A_3 with 5 elements. A_3 has two elements in common with both A_1 and A_2 and there is one element that is common to all three sets. If you sum the elements in all three sets then you will count elements 5, 7, and 8 twice and element 4 three times. Subtracting intersections between pairs of sets will zero out element 4 completely since it appears in three intersections. Element 4 is added back by adding the number of elements in the intersection of all three sets.

The general formula for three sets is:

$$|A_1 \cup A_2 \cup A_3| = |A_1| + |A_2| + |A_3| \qquad (1.26)$$
$$- |A_1 \cap A_2| - |A_1 \cap A_3|$$
$$- |A_2 \cap A_3| + |A_1 \cap A_2 \cap A_3|$$

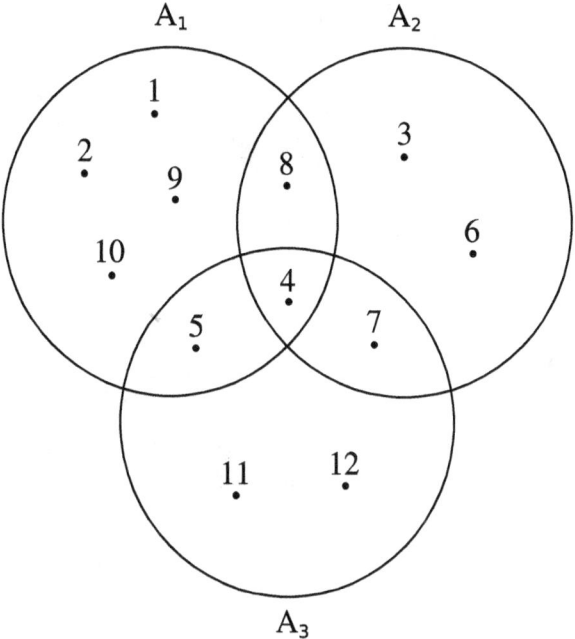

Figure 1.2: Three set example.

In general, to count the number of elements in the

union of n sets we have the following formula:

$$|A_1 \cup A_2 \cup \cdots A_n| = \tag{1.27}$$

$$\sum_{i=1}^{n} (-1)^{i-1} \sum_{j_1, j_2, \cdots, j_i} |A_{j_1} \cap A_{j_2} \cap \cdots A_{j_i}|$$

where the inner summation is over all i element subsets of $(1, 2, \ldots, n)$. To prove this formula you only have to show that any element in the union contributes a 1 to the summation on the right hand side. We will give a simple sketch of the proof.

Take an element x of the union that is a member of s of the sets. The $|A_i|$ terms will contribute a $\binom{s}{1} = s$ factor when counting the element. There will be $\binom{s}{2}$ factor from the $|A_i \cap A_j|$ terms, a $\binom{s}{3}$ factor from the $|A_i \cap A_j \cap A_k|$ terms and so on. The count for x is then equal to:

$$\binom{s}{1} - \binom{s}{2} + \binom{s}{3} - \cdots - (-1)^s \binom{s}{s} \tag{1.28}$$

You can see that this equation is always equal to 1 by comparing it to the expansion of $(y - 1)^s$

$$(y - 1)^s = y^s - \binom{s}{1} y^{s-1} + \binom{s}{2} y^{s-2} \tag{1.29}$$

$$- \binom{s}{3} y^{s-3} + \cdots + (-1)^s \binom{s}{s}$$

If you subtract this equation from 1 and set $y = 1$ you get equation 1.28. This means equation 1.28 is equal to $1 - (1 - 1)^s = 1$ which is what we set out to show.

1.6 Stirling Numbers of the Second Kind

How many ways can you partition a set of size n into k nonempty subsets? The question is equivalent to asking for the number of ways of putting n distinct balls into k identical urns so that each urn has at least one ball. The answer is given by the Stirling number of the second kind $S(n, k)$. We will derive a formula for $S(n, k)$ by using the inclusion exclusion principle. To begin, the number of ways to put n distinct balls into k distinct urns without restriction is k^n. The number of ways to put the balls into the urns so that at least i of the urns remain empty is $(k - i)^n$. There are $\binom{n}{i}$ ways that i of the urns can remain empty. So from the inclusion exclusion principle the number of ways to distribute the balls so that at least one urn remains empty is

$$\sum_{i=1}^{k-1} (-1)^{i-1} \binom{k}{i} (k - i)^n \qquad (1.30)$$

Now subtract this equation from k^n to get the number of ways to distribute n distinct balls into k distinct urns with at least one ball in each urn. The answer is

$$\sum_{i=0}^{k-1} (-1)^i \binom{k}{i} (k - i)^n \qquad (1.31)$$

This equation should equal $k!S(n,k)$ since in the case of the Stirling number the distribution is into urns that are identical. The equation for $S(n,k)$ must then be

$$S(n,k) = \frac{1}{k!}\sum_{i=0}^{k-1}(-1)^i\binom{k}{i}(k-i)^n \qquad (1.32)$$

An equivalent form of this equation is

$$S(n,k) = \frac{1}{k!}\sum_{i=1}^{k}(-1)^{k-i}\binom{k}{i}i^n \qquad (1.33)$$

The following formulas are for the first four k values.

$$S(n,1) = 1 \qquad (1.34)$$
$$S(n,2) = \frac{1}{2!}(2^n - 2) = 2^{n-1} - 1$$
$$S(n,3) = \frac{1}{3!}(3^n - 3\cdot 2^n + 3)$$
$$S(n,4) = \frac{1}{4!}(4^n - 4\cdot 3^n + 6\cdot 2^n - 4)$$

The numbers can also be calculated using the following recurrence

$$S(n,k) = S(n-1,k-1) + kS(n-1,k) \qquad (1.35)$$

with the initial conditions, $S(n,1) = 1$ for $n \geq 1$, and $S(n,n) = 1$ for all n. Given a value of $k \geq 0$ the generating function [1] for the numbers $S(n, k)$ is

$$\sum_{n \geq k} S(n, k)x^n = \qquad (1.36)$$

$$\frac{x^k}{(1 - x)(1 - 2x)(1 - 3x) \cdots (1 - kx)}$$

For a given value of n, summing $S(n, k)$ from $k = 1$ to $k = m$ gives the number of ways you can place n distinct balls into m identical bins with no restrictions. This is also the number of ways that a set of n elements can be partitioned into no more than m parts. The sum

$$\sum_{k=1}^{n} S(n, k) = B(n) \qquad (1.37)$$

is called a Bell number. It is equal to the total number of ways that a set of n elements can be partitioned into nonempty subsets.

---

[1]For a review of generating functions, see Herbert Wilf's book *generatingfunctionology*, available for free download here.

Problem 1. A furniture maker has 12 chair designs and 5 table designs. In how many ways can he make a chair and a table?

Answer. A chair design can be chosen from a set of 12, and a table design from a set of 5, so by the product rule, a chair and a table together can be chosen in $12 \cdot 5 = 60$ ways.

Problem 2. A friend shows me 5 novels and 7 math books, and allows me to choose one of each. In how many ways can I choose a pair?

Answer. There is a set of 5 and a set of 7, so by the product rule, a pair can be chosen in $5 \cdot 7 = 35$ ways.

Problem 3. If a nickel and a dime are tossed, how many ways can the coins fall?

Answer. Both the nickel and the dime can fall with heads up or tails up. So we are choosing from 2 sets each of size 2. The number of ways is therefore $2 \cdot 2 = 4$. If heads is H and tails is T, then all the ways is shown in table 2.1.

Problem 4. If 2 dice are thrown together, in how many ways can they fall?

Nickel	Dime
H	H
H	T
T	H
T	T

Table 2.1: Ways for Problem 3.

Answer. The first die can fall in 6 ways, and the second in 6 ways, so by the product rule they can fall together in $6 \cdot 6 = 36$ ways. All the ways are shown below.

1 and 1, 1 and 2, 1 and 3, 1 and 4, 1 and 5, 1 and 6,
2 and 1, 2 and 2, 2 and 3, 2 and 4, 2 and 5, 2 and 6,
3 and 1, 3 and 2, 3 and 3, 3 and 4, 3 and 5, 3 and 6,
4 and 1, 4 and 2, 4 and 3, 4 and 4, 4 and 5, 4 and 6,
5 and 1, 5 and 2, 5 and 3, 5 and 4, 5 and 5, 5 and 6,
6 and 1, 6 and 2, 6 and 3, 6 and 4, 6 and 5, 6 and 6.

Problem 5. In how many ways can 2 prizes be given to a class of 10 kids, without giving both to the same kid?

Answer. For the first prize we have a set of 10 kids to choose from, and for the second prize we have a set of 9 kids, so by the product rule, the number of ways that 2 prizes can be given without giving both to the same kid is $10 \cdot 9 = 90$.

Problem 6. In how many ways can 2 prizes be given

to a class of 10 kids, if it is allowed that both be given to the same kid?

Answer. The first prize can be given to one in a set of of 10 kids, and the second prize can also be given to one in a set of 10 kids, so by the product rule, the number of ways that 2 prizes can be given if it is allowed that both be given to the same kid is $10 \cdot 10 = 100$.

Problem 7. A friend shows me 5 physics books, 7 math books, and 10 history books, and allows me to choose 2 books, on the condition that they must not be both of the same subject. How many selections can I make?

Answer. I can apply the product rule to each possible subject pair, so I can choose a physics and a math book in $5 \cdot 7 = 35$ ways, a physics and a history book in $5 \cdot 10 = 50$ ways, and a math and a history book in $7 \cdot 10 = 70$ ways. And by the sum rule, I can make a total of $35 + 50 + 70 = 155$ selections.

Problem 8. With nine pairs of gloves, in how many ways can I choose a left hand glove and a right hand glove, so that the resulting pair is not a matching pair?

Answer. If I choose the left hand glove first, I have a set of 9 to choose from, then for the right hand glove I have a set of 8 to choose from, because the matching right hand glove is not included. So by the product rule, I can choose $9 \cdot 8 = 72$ possible pairs. Or... I can choose from 9 left hand gloves, and 9 right hand gloves, which gives $9 \cdot 9 = 81$ choices, but then I must subtract the 9 possible matching pairs, which leaves $81 - 9 = 72$ choices, as before.

Problem 9. Two people get into a bus that has 6 empty seats, in how many ways can they be seated?

Answer. The first person to board has a set of 6 seats to choose from, the second person has a set of 5 seats to choose from, so by the product rule, they can be seated in $6 \cdot 5 = 30$ different ways. The answer is also gotten by asking "How many permutations are there of 6 distinct items taken 2 at a time?" which is simply $\frac{6!}{(6-2)!} = 30$.

Problem 10. How many ways can we make a 2 letter word, out of an alphabet of 26 letters, if the 2 letters must be different?

Answer. For the first letter we have 26 letters to choose from, while for the second letter we have 25 letters to choose from, so the total number of possible words is $26 \cdot 25 = 650$ ways. The answer is also the number of permutations of 26 distinct things taken 2 at a time $= \frac{26!}{(26-2)!} = 650$.

Problem 11. In how many ways can we select a consonant and a vowel, if there are 20 consonants and 6 vowels?

Answer. The total number of consonant-vowel pairs we can select is $20 \cdot 6 = 120$.

Problem 12. In how many ways can we make a 2 letter word consisting of one consonant and one vowel?

Answer. From the previous answer, we have 120 possible consonant-vowel pairs. Each consonant-vowel pair can be arranged in 2 ways to form a word. So the total number of 2 letter words is $120 \cdot 2 = 240$.

Problem 13. There are 12 ladies and 10 gentlemen, of whom 3 ladies and 2 gentlemen are brothers and sisters, the rest being unrelated. How many ways can there be an unrelated marriage?

Answer. Neglecting relations, there are $12 \cdot 10 = 120$ possible marriages, but of those, there are $3 \cdot 2 = 6$ related pairs, so that leaves $120 - 6 = 114$ unrelated marriages.

Problem 14. In how many ways can the following set of letters be arranged in a row: $a, a, a, a, a, a, a, b, c$?

Answer. If we start by putting the 7 identical a's in a row with spaces between them:

$$a \ a \ a \ a \ a \ a \ a$$

then the b can be placed in any of the 6 spaces or on either end, making 8 possible places. When the b is placed, then the remaining letter c can be placed in any of 7 possible spaces or on either end, making 9 possible places. This results in a total of $8 \cdot 9 = 72$ possible arrangements.

Problem 15. The Local Galactic Congress consists of 8 humans, 13 Alpha Centaurians, 21 Lalandians, 11 Eridanians, 19 Aquarions, 10 Cygnians, and 15 Tau Cetians. In how many ways is it possible to form a committee of 7 members representing each of the 7 stellar life groups?

Answer. $8 \cdot 13 \cdot 21 \cdot 11 \cdot 19 \cdot 10 \cdot 15 = 68,468,400$.

Problem 16. Twenty people compete in a race with 3 prizes. In how many ways can the prizes be awarded?

Answer. The first prize can be awarded 20 ways, the second 19 ways, and the third 18 ways, so the total number of ways is $20 \cdot 19 \cdot 18 = 6,840$. Or, it is simply the number of permutations of 20 objects taken 3 at a time $= \frac{20!}{(20-3)!} = 6,840$.

Problem 17. How many ways can 4 letters be put into 4 envelopes, with only one letter per envelope?

Answer. For the first letter there are 4 envelopes to choose from, for the second letter there are 3 envelopes, for the third, 2 envelopes, and for the fourth, one. The total number of ways is then $4 \cdot 3 \cdot 2 \cdot 1 = 24$. It can also be stated as the number of permutations of 4 distinct items taken 4 at a time $= \frac{4!}{(4-4)!} = 24$.

Problem 18. How many ways can you sum a silver dollar, a quarter, a dime, a nickel, and a penny?

Answer. Each coin can either be summed or not, and there are 5 coins, so the ways you can sum them is $2^5 = 32$. But this includes the case of not summing any coin, which isn't really summing them, so the answer is 31.

Problem 19. There are 20 candidates for an office, and 7 voters. How many ways can the votes be made?

Answer. Each of the 7 voters can choose one of 20 candidates, so by the product rule, the number of ways is $20^7 = 1,280,000,000$

Problem 20. I have 6 letters to be delivered to different parts of town, and two boys offer their services to deliver them. In how many different ways can I choose to send the letters?

Answer. I can choose to give the first letter to one of either boys, the second letter to one of either boys, and so on for all 6 letters. Therefore there are 2 choices for each letter, so the number of ways is $2^6 = 64$.

Problem 21. In how many ways can 6 different things be divided between 2 boys?

Answer. This question appears to be identical to the previous, so the answer should be 64. But the answer to the previous question included the case of one boy delivering all the letters, and the other boy having none to deliver, which isn't really a way to divide the letters between the 2 boys. Therefore the answer is $64 - 2 = 62$.

Problem 22. In how many ways can six different things be divided into two parcels?

Answer. This question seems identical to the previous, but that question made a distinction between the two boys, whereas with parcels, we don't. For example the first boy could have 4 things, and the second boy 2. This was distinguished from the first boy having 2 things, and the second boy 4. With parcels there is no such distinction, so the answer is not 62 like previously, but half of that, 31.

Problem 23. In how many ways can the following fruits be divided between 2 people: *fig*, *fig*, *fig*, *fig*, *date*, *date*, *date*, *kiwi*, *kiwi*, *lime*?

Answer. Considering only how the 4 figs can be divided, one person can get none, or just one, or 2, 3, or 4. So there are 5 ways to divide the 4 figs. Similarly there are 4 ways to divide the 3 dates, 3 ways to divide the 2 kiwis, and 2 ways to divide the single lime. So all the fruits can be divided in $5 \cdot 4 \cdot 3 \cdot 2 = 120$ ways. But this includes the case of one person getting all the fruits, and the other getting none, which isn't really a way to divide them, so excluding those 2 cases, the answer is 118.

Problem 24. In the ordinary system of notation, how many numbers are there which consist of five digits?

Answer. The leftmost digit can have any of the 10 numerals except for 0, while the remaining 4 digits can have any of the 10 numerals. So the total

number of numbers which consist of 5 digits is
$9 \cdot 10 \cdot 10 \cdot 10 \cdot 10 = 90,000$.

Problem 25. If a password consists of 4 letters, each letter chosen from 26 possibilities, how many different possible passwords are there?

Answer. $26^4 = 456,976$.

Problem 26. In how many ways can we arrange 6 statues in 6 niches, one in each?

Answer. For the first statue there are 6 niches to choose from, for the second statue, 5 niches, for the third statue, 4 niches, for the fourth statue, 3 niches, for the fifth statue, 2 niches, and for the sixth statue 1 niche. So the total number of ways we can arrange 6 statues in 6 niches is
$6! = 6 \cdot 5 \cdot 4 \cdot 3 \cdot 2 \cdot 1 = 720$.

Problem 27. In how many ways can 12 ladies and 12 gentlemen form themselves into couples for a dance?

Answer. The first gentleman has 12 ladies to choose from, the second gentleman has 11 ladies to choose from, the third gentleman has 10 ladies to choose from, and continuing until the twelfth gentleman has one lady to choose. Therefore 12 ladies and

12 gentlemen can form themselves into $12! = 12 \cdot 11 \cdot 10 \ldots 3 \cdot 2 \cdot 1 = 479,001,600$ couples.

Problem 28. With 12 ladies and 12 gentlemen in a ballroom, in how many ways can they take their places for a contra dance (couples dance in 2 facing lines)?

Answer. From the previous question, there are 12! different possible couples, and the 12 couples can be arranged in a line in 12! different ways. So the total number of ways they can take their places for a contra dance is $12! \cdot 12! = 229,442,532,802,560,000$. Alternatively, the gentlemen can arrange themselves in a line in 12! different ways, and the ladies can arrange themselves in a line in 12! different ways. So the total number of ways they can take their places is $12! \cdot 12!$ as we just found.

Problem 29. In how many different ways can the letters
a, b, a, d, e, f be arranged so that each arrangement begins with ab?

Answer. If each arrangement begins with ab, it's as if we simple removed those 2 letters from the arrangements, so that we only have the remaining 4 letters: a, d, e, f, and 4 items can be arranged in $4! = 24$ ways.

Problem 30. A shelf holds 5 biology books, 6 history, and 8 math books. In how many ways can the 19 books be arranged, keeping all the biology together, all the history together, and all the math together?

Answer. The 5 biology books can be arranged in $5! = 120$ ways, the 6 history books in $6! = 720$ ways, and the 8 math books in $8! = 4,320$ ways. The 3 different categories of books can be arranged in $3! = 6$ ways. So the total number of possible arrangements is the product of these ways: $120 \cdot 720 \cdot 40,320 \cdot 6 = 20,901,888,000$ ways.

Problem 31. In how many ways can the same books be arranged indiscriminately on the shelf?

Answer. $19! = 121,645,100,408,832,000$ ways.

Problem 32. A table is set for 8 people. How many ways can they be seated?

Answer. $8! = 40,320$ ways.

Problem 33. How many ways can 8 children form themselves in a ring to dance around a maypole?

Answer. In this case, relative position is all that matters. Take for example the 4 arrangements, where the 8 children are represented by letters A through H.

	A	B			C	D			D	E			A	H	
H				C	B		E	C			F	B				G
G				D	A		F	B			G	C				F
	F	E			H	G			A	H			D	E	

If we first place A, and all the others relative to it, we see that the first 3 arrangements above are identical, but the last is different (having its circularity reversed). Since relative position is all that matters, then the number of possible arrangements are to be counted after the first is placed, leaving $7! = 5,040$ ways.

Problem 34. In how many ways can eight beads be strung on an elastic band to form a bracelet?

Answer. This question appears to be identical to the last, but there is a difference. For the 4 arrangements shown in the previous question, the rightmost arrangement is different for children around a maypole, but it's the same for a bracelet, since it only represents a rotation of the bracelet, and not a difference in its construction. By this reasoning, for a bracelet there are only half as many unique arrangements as for children around a maypole. This makes $\frac{7!}{2} = 2,520$ ways.

Problem 35. How many 3 letter words can be made from an alphabet of 26 letters, with each letter used only once?

Answer. $26 \cdot 25 \cdot 24 = 15,600$ words.

Problem 36. How many 4 letter words?

Answer. $26 \cdot 25 \cdot 24 \cdot 23 = 358,800$ words.

Problem 37. How many 8 letter words?

Answer. $26 \cdot 25 \cdot 24 \cdot 23 \cdot 22 \cdot 21 \cdot 20 \cdot 19 = 62,990,928,000$ words.

Problem 38. 4 flags are to be raised on one mast, with 20 different flags to choose from. How many different ways can the flags be arranged?

Answer. $20 \cdot 19 \cdot 18 \cdot 17 = 116,280$ ways. The answer would be the same if there were 4 different masts for the 4 flags, since there are still only 4 positions.

Problem 39. A boat with 8 oars has to be manned by a club with 50 members. How many ways can we arrange the crew?

Answer. $50 \cdot 49 \cdot 48 \cdot 47 \cdot 46 \cdot 45 \cdot 44 \cdot 43 = 21,646,947,168,000$ ways.

Problem 40. In how many ways can 2 people divide 30 books between them, with one having twice as many as the other?

Answer. This is the number of combinations of 30 taken 20 at a time, or $\binom{30}{20} = \frac{30!}{20! \cdot 10!} = 30,045,015$ ways.

Problem 41. There are 8 men to row an 8 oared boat, but 2 of them can only row on stroke side, and one of them only on bow side, with the rest able to row on either side. How many possible arrangements of the men are there?

Answer. Of the 5 remaining men who can row on either side, 2 of them are needed for stroke side, and 3 on bow side. These 5 men can therefore be arranged in $\binom{5}{2} = \frac{5!}{2! \cdot 3!} = 10$ ways. The 4 men on stroke side can be arranged in $4! = 24$ ways, and the 4 men on bow side can similarly be arranged in 24 ways. So the total number of possible arrangements is $10 \cdot 24 \cdot 24 = 5,760$ ways.

Problem 42. In how many ways can 3 kids divide 12 oranges, the oranges all being of different sizes?

Answer.

$$\binom{12}{4,4,4}$$
$$= \frac{12!}{4! \cdot 4! \cdot 4!}$$
$$= 34,650$$

ways.

Problem 43. In how many ways can they divide them so that the eldest gets 5, the middle 4, and the youngest 3?

Answer.

$$\binom{12}{5, 4, 3}$$
$$= \frac{12!}{5! \cdot 4! \cdot 3!}$$
$$= 27,720$$

ways.

Problem 44. If there are 15 identical apples, 20 identical pears, and 25 identical oranges, in how many ways can 60 kids take one each?

Answer.

$$\binom{60}{15, 20, 25}$$
$$= \frac{60!}{15! \cdot 20! \cdot 25!}$$
$$= 168,618,391,667,123,831,595,882,720$$
$$\approx 1.7 \text{x} 10^{26}$$

ways.

Problem 45. Using 6 dice, in how many ways can you throw 2 sixes, 3 fives, and a one?

Answer.

$$\binom{6}{2,3,1} = \frac{6!}{2! \cdot 3! \cdot 1!} = 60$$

ways.

Problem 46. In how many ways can a deck of 52 cards be divided among 4 players so that each gets 13 cards?

Answer.

$$\binom{52}{13,13,13,13}$$
$$= \frac{52!}{13! \cdot 13! \cdot 13! \cdot 13!}$$
$$= 53,644,737,765,488,792,839,237,440,000$$
$$\approx 5.4 \text{x} 10^{28}$$

ways.

Problem 47. In how many ways can a deck of 52 cards be divided into 4 groups of 13 cards?

Answer. This question is different from the previous only by the fact that the 4 players in the previous question were distinguishable, whereas the 4 groups in this problem are not. The consequence is that the answer to the previous question must be divided by 4!. So we have

$$\binom{52}{13,13,13,13} \cdot \frac{1}{4!}$$
$$= \frac{52!}{13! \cdot 13! \cdot 13! \cdot 13! \cdot 4!}$$
$$= 2,235,197,406,895,366,368,301,560,000$$
$$\approx 2.2 \text{x} 10^{27}$$

ways.

Problem 48. In how many ways can 12 books be divided into 3 groups, with one group having 3 members, another 4 members, and the last 5 members?

Answer. This question is different from the previous by the fact that the 3 groups in this question have different numbers of members, with the consequence that the groups are distinguishable, therefore the number of ways is $\binom{12}{3,4,5} = 27,720$.

Problem 49. In how many ways can we arrange the letters of the word *indivisibility*?

Answer. If all 14 letters of this word were unique, the answer would be 14!. But there are 6 identical letters (the i's). The result is that 14! is larger than the answer by a factor of 6!, so the number of ways is $\frac{14!}{6!} = 121,080,960$.

Problem 50. In how many ways can we arrange the letters of the word *parallelepiped*?

Answer. By similar reasoning to the previous problem, we have 14 letters total, but the duplicates are 3 p's, 2 a's, 3 l's, and 3 e's, so the number of ways is $\frac{14!}{3! \cdot 2! \cdot 3! \cdot 3!} = 201,801,600$.

Problem 51. In how many ways can we arrange the letters of the word *llangollen*?

Answer. $\frac{10!}{4! \cdot 2!} = 75,600$ ways.

Problem 52. Out of 100 things, how many ways can 3 things be selected?

Answer. $\binom{100}{3} = \frac{100!}{3! \cdot 97!} = 161,700$ ways.

Problem 53. From a basket of twenty pears, at 3 pears for a dollar, how many ways can you select 6 dollars worth?

Answer. Our 18 pears can be selected in $\binom{20}{18} = \frac{20!}{18! \cdot 2!} = \frac{20 \cdot 19}{2} = 190$ ways.

Problem 54. In how many ways can we make the same choice, if we always take the largest pear?

Answer. Always choosing the largest pear, there are 17 additional pears to choose from the 19 remaining, so the number of ways is $\binom{19}{17} = \frac{19!}{17! \cdot 2!} = \frac{19 \cdot 18}{2} = 171$.

Problem 55. In how many ways can we make the same choice without taking the smallest pear?

Answer. Taking the smallest pear out of the running, there are 18 to choose from the 19 remaining, so the number of ways is $\binom{19}{18} = 19$.

Problem 56. In how many ways can we make the same choice by both included the largest pear, and not including the smallest pear?

Answer. After taking the largest pear, and removing the smallest pear from the running, there are 17 pears to choose from the 18 remaining. This results in $\binom{18}{17} = \frac{18!}{17!} = 18$ ways.

Problem 57. Out of 42 liberals and 50 conservatives, how many choices are there in selecting a committee of 4 liberals and 4 conservatives?

Answer. $\binom{42}{4} \cdot \binom{50}{4} = \frac{42 \cdot 41 \cdot 40 \cdot 39}{4!} \cdot \frac{50 \cdot 49 \cdot 48 \cdot 47}{4!} = 111,930 \cdot 230,300 = 25,777,479,000$ ways.

Problem 58. A company of volunteers consists of a captain, a lieutenant, an ensign, and 80 rank and file. In how many ways can 10 soldiers be selected so that the captain is always included?

Answer. Always choosing the captain, then there are 9 remaining to choose from 82 soldiers. So the number of ways is $\binom{82}{9} = \frac{82!}{73! \cdot 9!} = 293,052,087,900$.

Problem 59. How many ways can 10 soldiers be selected so that at least one officer is included?

Answer. There are $\binom{83}{10}$ ways to choose 10 soldiers from the 3 officers and the 80 rank and file. There are $\binom{80}{10}$ ways to choose 10 soldiers from the 80 rank and file. The difference $\binom{83}{10} - \binom{80}{10} = 785,840,219,450$ is the answer. We may also get the same result as follows. There are $3 \cdot \binom{80}{9}$ ways to choose one officer and the remaining 9 soldiers from rank and file. There are $3 \cdot \binom{80}{8}$ ways to choose 2 officers and the remaining 8 soldiers from rank and file. And there are $1 \cdot \binom{80}{7}$ ways to choose 3 officers and the remaining 7 soldiers from rank and file. The sum of these ways gives the same answer as above: $3 \cdot \binom{80}{9} + 3 \cdot \binom{80}{8} + 1 \cdot \binom{80}{7} = 785,840,219,450$.

Problem 60. How many ways can 10 soldiers be selected so that exactly one officer is included?

Answer. There are 3 ways to choose one officer, and $\binom{80}{9}$ ways to choose the 9 remaining soldiers from rank and file, therefore the number of ways is $3 \cdot \binom{80}{9} = 695,700,891,600$.

Problem 61. There are 15 candidates for admission into a society with 2 vacancies. There are 7 electors and each can either vote for one or 2 candidates. How many ways can the votes be made?

Answer. Each voter can choose a single candidate 15 ways, and 2 candidates $\binom{15}{2}$ ways, for a total of 120 ways. There are 7 voters, so the total number of ways the votes can be made is $120^7 = 358,318,080,000,000$.

Problem 62. Out of 20 men and 6 women, in how many ways can we choose 3 men and 3 women?

Answer. $\binom{20}{3} \cdot \binom{6}{3} = 1,140 \cdot 20 = 22,800$ ways.

Problem 63. Out of 20 men and 6 women, in how many ways can we fill 6 offices, with 3 requiring men and 3 requiring women?

Answer. This question is different from the previous in that the order of selection is important, so the number of ways is $\frac{20!}{(20-3)!} \cdot \frac{6!}{(6-3)!} = 20 \cdot 19 \cdot 18 \cdot 6 \cdot 5 \cdot 4 = 820,800$.

Problem 64. Using 20 consonants and 6 vowels, how many ways can we make a word with 3 different consonants and 3 different vowels?

Answer. We can choose 3 different consonants from 20 in $\binom{20}{3}$ ways, we can choose 3 different vowels from 6 in $\binom{6}{3}$ ways, and the number of permutations of the chosen 6 letters is 6!, so the total number of ways to make a word with 3 consonants and 3 vowels is the product of these ways: $\binom{20}{3} \cdot \binom{6}{3} \cdot 6! = \frac{20 \cdot 19 \cdot 18}{6} \cdot \frac{6 \cdot 5 \cdot 4}{6} \cdot 6 \cdot 5 \cdot 4 \cdot 3 \cdot 2 = 1,140 \cdot 20 \cdot 720 = 16,416,000.$

Problem 65. From the 26 letters of the alphabet, how many ways can we make a word with 4 different letters, where one of the letters must be a?

Answer. If one of the letters must be a, then there are 3 more letters to choose from the remaining 25 letters. The 3 letters can be chosen from the 25 in $\binom{25}{3} = \frac{25 \cdot 24 \cdot 23}{6} = 2,300$ ways. The 4 letters (including the a) can be arranged in $4! = 24$ ways, so the total number of ways we can make a word with 4 different letters, that includes an a, is $2,300 \cdot 24 = 55,200$. Alternatively, we can solve this problem by finding the number of 4 letter words, where the letters are different but there is no restriction on the a, as $26 \cdot 25 \cdot 24 \cdot 23 = 358,800$ words. The number of letters in these 4 letter words is $4 \cdot 358,800 = 1,435,200$. Now we note that since there was no bias in picking the letters, the number of a's in these 1,435,200 letters is $1,435,200/26 = 55,200$. This is also the number of words with a's in them, which matches the answer gotten previously.

Problem 66. From the 26 letters of the alphabet, how many ways can we make a 4 letter word, with 2 of them being a and b, and all of them being different?

Answer. Having already chosen a and b, we have 24 letters to choose 2 letters from, which can be done in $\binom{24}{2}$ ways. The number of ways to order 4 letters out of 4 is 4!. So the total number of ways is $\binom{24}{2} \cdot 4! = 276 \cdot 24 = 6,624$.

Problem 67. From 20 consonants and 6 vowels, how many ways can we make a 5 letter word consisting of 2 different consonants and 3 different vowels, one of which must be a?

Answer. The 2 consonants can be chosen in $\binom{20}{2}$ ways, the 2 vowels (besides the a) can be chosen in $\binom{5}{2}$ ways, and the 5 letters can be ordered in 5! ways, so the total number of ways to make the 5 letter word are $\binom{20}{2} \cdot \binom{5}{2} \cdot 5! = 190 \cdot 10 \cdot 120 = 228,000$.

Problem 68. There are 10 jobs that need to be filled, 4 of which must be filled by men, 3 by women, and the remaining 3 by either. If there are 20 men and 6 women available, in how many ways can we fill the jobs?

Answer. The 20 men can fill the 4 jobs in $\frac{20!}{(20-4)!}$ ways, the 6 women can fill the 3 jobs in $\frac{6!}{(6-3)!}$ ways. There are 16 men and 3 women still available to fill the 3 remaining jobs, which can be done in $\frac{19!}{(19-3)!}$ ways. So the total number of ways to fill the 10 jobs is $\frac{20!}{(20-4)!} \cdot \frac{6!}{(6-3)!} \cdot \frac{19!}{(19-3)!} = 116,280 \cdot 120 \cdot 5,814 = 81,126,230,400.$

Problem 69. From a large number of nickels, dimes, and quarters, in how many ways can 4 coins be selected?

Answer. This question is equivalent to asking "In how many ways can I put 4 items into 3 urns?". We can represent this with stars (items) and bars (urn boundaries). For example, one way to put 4 items into 3 urns is:

$$* \mid * \; * \mid *$$

This represents one item in the first urn, 2 items in the second urn, and one item in the third. The answer is the total number of ways of arranging these 6 items, with 4 of one kind (stars) and 2 of another (bars) in a line, or $\binom{6}{2} = \frac{6!}{2! \cdot 4!} = \frac{6 \cdot 5}{2} = 15$ ways. All the ways are shown below:

n n n n, n n d d, n d d d, n q q q, d d q q,
n n n d, n n d q, n d d q, d d d d, d q q q,
n n n q, n n q q, n d q q, d d d q, q q q q.

Problem 70. How many ways can we fill 3 glasses with 5 wines, without mixing?

Answer. Similar to the previous problem, this question can be posed as "In how many ways can I put 3 items into 5 urns?". Thinking in terms of "stars and bars" (see previous problem) the answer is the total number of ways of arranging 7 items, with 3 of one kind (stars) and 4 of another (bars) in a line, or $\binom{7}{3} = \frac{7!}{3! \cdot 4!} = \frac{7 \cdot 6 \cdot 5}{6} = 35$ ways. If we represent the 5 types of wine with vowels (a, e, i, o, u), then the 35 ways are shown below:

$$aaa, \quad aee, \quad aio, \quad eee, \quad eio, \quad iii, \quad iuu,$$
$$aae, \quad aei, \quad aiu, \quad eei, \quad eiu, \quad iio, \quad ooo,$$
$$aai, \quad aeo, \quad aoo, \quad eeo, \quad eoo, \quad iiu, \quad oou,$$
$$aao, \quad aeu, \quad aou, \quad eeu, \quad eou, \quad ioo, \quad ouu,$$
$$aau, \quad aii, \quad auu, \quad eii, \quad euu, \quad iou, \quad uuu.$$

This answer assumes the glasses are indistinguishable, so that aei is the same as aie, eai, eia, iae, and iea. If the glasses are unique, then each can be filled in 5 ways, so the number of ways would be $5^3 = 125$.

Problem 71. In how many ways can a dozen marbles be selected in a shop where they sell 5 kinds of marbles?

Answer. Similar to the previous 2 problems, this question is equivalent to asking "In how many ways can I put 12 marbles into 5 urns?". So there are 12 stars and 4 bars, meaning the total number of ways to arrange 16 items with 12 of one kind and 4 of another is $\binom{16}{12} = \frac{16 \cdot 15 \cdot 14 \cdot 13}{4!} = 1,820$.

Problem 72. How many dominoes are there in a set numbered from double blank to double 9?

Answer. Each domino contains a selection of 2 numbers from 10 possibilities: 0, 1, 2, 3, 4, 5, 6, 7, 8, 9. Similar to the previous 3 problems, this question is equivalent to asking "In how many ways can I put 2 items into 10 urns?". So there are 2 stars and 9 bars, meaning the total number of ways to arrange 11 items with 2 of one kind and 9 of another is $\binom{11}{2} = \frac{11 \cdot 10}{2} = 55$.

Problem 73. In how many ways can an arrangement of 4 letters be made from letters of the words *choice and chance*?

Answer. There are 15 letters to choose from, with 8 unique ones. Duplicates are indicated in the following table. The ways the 4 letters can be

c	h	a	n	e	o	i	d
4	2	2	2	2	1	1	1

chosen falls into 5 categories:

1. Four all the same.

2. Three the same, one different.

3. Two the same, the other two the same.

4. Two the same, the other two different.

5. All four different.

Category 1 allows only one way, 4 c's. Category 2 requires 3 c's and one of the other 7 letters, with the other letter occupying one of 4 positions, so there are $7 \cdot 4 = 28$ ways. For category 3 we can choose the 2 pairs in $\binom{5}{2} = 10$ ways, and the 2 pairs can be arranged in $\binom{4}{2} = 6$ ways, so there are $10 \cdot 6 = 60$ ways. For category 4, we can choose the pair in 5 ways, the other 2 in $\binom{7}{2} = 21$ ways, and the 4 letters can be arranged in $\binom{4}{2} = 12$ ways for a total of $5 \cdot 21 \cdot 12 = 1,260$ ways. For category 5, we have to choose 4 unique letters from 8, so the number of ways is $8 \cdot 7 \cdot 6 \cdot 5 = 1,680$. The total number of arrangements of 4 letters from the given 15 is then the sum of the arrangements from each category: $1 + 28 + 60 + 1,260 + 1,680 = 3,029$.

Problem 74. In how many ways can an arrangement be made of 3 things chosen from 15 things where 5 are of one type, 4 of another type, 3 of another type, and the last 3 of another type?

Answer. The ways the 3 things can be chosen falls into 3 categories:

1. Three all the same.

2. Two the same, one different.

3. All three different.

For category 1, there are 4 ways to choose all 3 the same. For category 2, there are 4 ways to choose 2 the same, and 3 ways to choose one different, and $\binom{3}{2} = 3$ ways to arrange what was chosen, for a total of $4 \cdot 3 \cdot 3 = 36$ ways. For category 3, there are $4 \cdot 3 \cdot 2 = 24$ ways to choose all 3 different things. The combined number of ways to choose 3 things from the 15 described things is then $4 + 36 + 24 = 64$.

Problem 75. In how many ways can an arrangement be made of 5 things chosen from the 15 things described in the previous problem?

Answer. The ways the 5 things can be chosen falls into 6 categories:

1. Five all the same.

2. Four the same, one different.

3. Three the same, another two the same.

4. Three the same, two different.

5. Two the same, another two the same, one different.

6. Two the same, three different.

For category 1, there is only one way to choose. For category 2, there are 6 ways to choose with 5 ways to arrange for a total of $6 \cdot 5 = 30$ ways. For category 3, there are 12 ways to choose, with $\binom{5}{3,2} = 10$ arrangements, so there are $12 \cdot 10 = 120$ ways. For category 4, there are 12 ways to choose, with $\binom{5}{3,1,1} = 20$ arrangements, so there are $12 \cdot 20 = 240$ ways. For category 5, there are 12 ways to choose, with $\binom{5}{2,2,1} = 30$ arrangements, so there are $12 \cdot 30 = 360$ ways. For category 6, there are 4 ways to choose, with $\binom{5}{2,1,1,1} = 60$ arrangements, so there are $4 \cdot 60 = 240$ ways. The combined number of ways to choose 5 things from the 15 described things is then $1+30+120+240+360+240 = 991$.

Problem 76. How many four letter words can you make with an alphabet of 20 letters if repetitions are allowed?

Answer. There are 20 choices for each of the 4 letters so the number of possible words is: $20 \cdot 20 \cdot 20 \cdot 20 = 20^4 = 160,000$.

Problem 77. How many five letter words can you make with the 5 vowels with repetitions allowed?

Answer. There are 5 choices for each of the 5 letters so the number of possible words is: $5 \cdot 5 \cdot 5 \cdot 5 \cdot 5 = 5^5 = 3,125$.

Problem 78. Using the ten decimal digits 0, 1, 2, 3, 4, 5, 6, 7, 8, 9 how many six digit numbers can you make if leading zeros are not allowed?

Answer. Not allowing leading zeros means that the first digit must be something other than a 0 so there are 9 choices for the first digit. There are 10 choices each for the remaining digits. The number of possible six digit numbers is then: $9 \cdot 10 \cdot 10 \cdot 10 \cdot 10 \cdot 10 = 9 \cdot 10^5 = 900,000$.

Problem 79. There are twelve books for sale. In how many ways can you purchase one or more books?

Answer. If you purchase only one book then you have 12 choices. If you purchase two books then the number of ways of choosing 2 books from 12 is $\binom{12}{2}$. In general if you decide to purchase k books then the number of ways you can choose the k books is $\binom{12}{k}$. The sum of all the possibilities is:

$$\binom{12}{1} + \binom{12}{2} + \cdots + \binom{12}{12}$$

The easy way to do this sum is to remember that these binomial coefficients appear in the expansion:

$$(1+x)^{12} = \binom{12}{0} + \binom{12}{1}x + \binom{12}{2}x^2 \cdots \binom{12}{12}x^{12}$$

If you set $x = 1$ in this equation then the sum on the right is the same as the sum we are trying to calculate except for the $\binom{12}{0} = 1$ term. On the left of the equation we get 2^{12} so subtracting 1 from this gives us the answer: $2^{12} - 1 = 4,095$. In general this question is asking how many subsets of n elements can be formed. If you include the possibility of the empty subset then the answer is 2^n. If you exclude the empty subset then the answer is $2^n - 1$.

Problem 80. How many different weight values can you make by combining five individual weights?

Answer. The answer will depend on the exact values of the five weights. In general they can be combined to give $2^5 - 1 = 31$ weight totals but it may be that not all totals are different. If for example the weights are $2, 4, 6, 8, 10$ then the two different combinations $2 + 8$ and $4 + 6$ both give a value of 10. If the weight values are $1, 2, 4, 8, 16$ then all the possible combinations will be unique and every integer weight value from 1 to 31 will be produced by a combination. In this case the different combinations can be represented as five digit binary numbers. For example 01101 represents the combination $1 + 4 + 8 = 13$. As we know, the number of five digit binary numbers, excluding zero, is $2^5 - 1 = 31$.

Problem 81. A weight w has to be measured on a balance using five counter weights. The counter weights can be placed on either side of the balance. That is they can be put on the same side as the weight to be measured or on the opposite side. How many different w weights can be balanced by the five counter weights.

Answer. In general each counter weight can be put on the side of the weight w, on the opposite side or left off the scale. There are 3 options for each of the five counter weights so there are $3^5 - 1 = 242$ ways they can be used (excluding the case where all 5 are left off the scale). But only half of these combinations will allow a positive weight w to be measured. To see this let w_1, w_2, w_3, w_4, w_5 be the five counter weights then w can be expressed as the sum of the weights placed on the opposite side of the balance minus the sum of the weights placed on the same side. For example if w_1 and w_2 are on the same side as w, and w_5 is on the opposite side then we must have $w + w_1 + w_2 = w_5$ or $w = w_5 - w_1 - w_2$. You can now multiply this last equation by -1 which amounts to switching the counter weights from one side to the other. But then the measured weight w is negative. Therefor for every combination that allows a positive weight to be measured there will be a combination resulting in a negative measured weight. If these latter combinations are excluded then there are only $\frac{242}{2} = 121$ valid combinations. Whether all these combinations measure a unique weight w will depend on the exact values of the counter weights. The weight set $1, 3, 9, 27, 81$ will allow 121 unique weights to be measured. You can see this by noting that there is a one to one correspondence between combinations of the counter weights and 5 digit trinary numbers. Each combination of counter weights can be written as:

$$w = c_1 w_1 + c_2 w_2 + c_3 w_3 + c_4 w_4 + c_5 w_5 \quad (2.1)$$

where $c_i = 1$ if w_i is placed opposite to w, $c_i = -1$ if w_i is placed on the same side as w and $c_i = 0$ if w_i is left off the scale. A 5 digit trinary number on the other hand can be written as:

$$t_0 + t_1 3 + t_2 3^2 + t_3 3^3 + t_4 3^4 \quad (2.2)$$

where each trinary digit can be 0, 1, or 2. If $w_i = 3^i$ and $c_i = t_{i-1} - 1$ then there is a one to one correspondence between the unique values of both equations.

Problem 82. In how many ways can two booksellers split an inventory of 200 copies of Tarzan, 250 copies of Robinson Crusoe, 150 copies of Don Quixote, and 100 copies of Marco Polo?

Answer. One of the booksellers can take anywhere from 0 to 200 copies of Tarzan with the other taking the rest so there are 201 ways they can split Tarzan. Similarly there are 251 ways to split Robinson Crusoe, 151 ways to split Don Quixote, and 101 ways to split Marco Polo. The total number of ways they can split the books is then: $201 \cdot 251 \cdot 151 \cdot 101 = 769,428,201$. In one of these ways the first bookseller gets none of the books and in another one of the ways the second bookseller gets none of the books. If you exclude these two possibilities then the number of ways they can split the books is equal to $769,428,201 - 2 = 769,428,199$.

Exercise 1. If there are 6 routes from Yellowstone National Park to Glacier National Park, and 4 routes from Glacier National Park to Banff National Park, how many ways can you go from Yellowstone to Banff via Glacier?

Answer. By the product rule, there are $6 \cdot 4 = 24$ ways.

Exercise 2. Two groups of 100 students compete in a math olympiad. In how many ways can a champion be chosen from each group?

Answer. $100 \cdot 100 = 10,000$ ways.

Exercise 3. For a dog and pony show, we must choose one dog from 7, and one pony from 4. In how many ways can we choose a dog and a pony?

Answer. $7 \cdot 4 = 28$ ways.

Exercise 4. How many ways can we choose a consonant and a vowel from letters of the word *almost*?

Answer. There are 2 vowels and 4 consonants, so we can choose a consonant and a vowel in $2 \cdot 4 = 8$ ways.

Exercise 5. How many ways can we choose a consonant and a vowel from letters of the word *orange*?

Answer. There are 3 vowels and 3 consonants, so we can choose a consonant and a vowel in $3 \cdot 3 = 9$ ways.

Exercise 6. A die of 6 faces and a teetotum of 8 faces are thrown. In how many ways can they fall?

Answer. $6 \cdot 8 = 48$ ways.

Exercise 7. There are 3 major routes to the top of Longs Peak: the Keyhole route, the Loft route, and Keplinger's Couloir. In how many ways can a person go up and down?

Answer. $3 \cdot 3 = 9$ ways.

Exercise 8. With 20 forks and 24 spoons available, how many ways can a man choose a fork and a spoon? Then how can another man take another fork and spoon?

Answer. The first man can choose a fork and spoon in $20 \cdot 24 = 480$ ways, and the second man can choose a fork and spoon in $19 \cdot 23 = 437$ ways.

Exercise 9. From a list of 9 Corvidae birds, and 48 Columbidae birds, in how many ways can we choose an example of each?

Answer. $9 \cdot 48 = 432$ ways.

Exercise 10. From 12 masculine words, 9 feminine, and 10 neuter, in how many ways can we choose an example of each?

Answer. $12 \cdot 9 \cdot 10 = 1,080$ ways.

Exercise 11. With 6 pairs of unique gloves, how many ways can you take a left hand glove and a right hand glove that are unmatched?

Answer. A left hand glove can be chosen in 6 ways, and an unmatched right hand glove can be chosen in 5 ways, so there are a total of $6 \cdot 5 = 30$ ways.

Exercise 12. From 5 bibles and 12 prayer books, how many ways can I select a bible and a prayer book?

Answer. $5 \cdot 12 = 60$ ways.

Exercise 13. From 7 unicycles, 5 bicycles, and 8 skateboards, how many ways can I select one of each?

Answer. $7 \cdot 5 \cdot 8 = 280$ ways.

Exercise 14. There are 6 bibles, 3 prayer books, and 4 hymn books, in addition to 5 volumes of a bible and prayer book bound together, and 7 volumes of a prayer and hymn book bound together. In how many ways can I select a bible, a prayer book, and a hymn book?

Answer. As singles, there are $6 \cdot 3 \cdot 4 = 72$ ways to select, using a bible + prayer book bound volume there are $5 \cdot 4 = 20$ ways, and using a prayer + hymn book bound volume there are $7 \cdot 6 = 42$ ways, for a total of $72 + 20 + 42 = 134$ ways.

Exercise 15. If there were also 3 volumes of a bible and hymn book bound together, how many ways would there now be?

Answer. In addition to the above 134 ways, I could also use a bible + hymn book bound volume for $3 \cdot 3 = 9$ more ways, for a total of $134 + 9 = 143$ ways.

Exercise 16. A bowl holds 12 mangos and 10 grapefruits, Spike will choose a mango *or* a grapefruit, and then Spud will choose a mango *and* a grapefruit. Show that if Spike chooses a mango, then Spud will have more ways to choose than if Spike chose a grapefruit.

Answer. If Spike chooses a mango, then Spud will have $11 \cdot 10 = 110$ ways to choose a mango and a grapefruit. On the other hand, if Spike chooses a grapefruit then Spud will have $12 \cdot 9 = 108$

ways to choose. Therefore Spud has more ways
to choose, if Spike chooses a mango.

Exercise 17. In how many different permutations can
8 bells be rung? And in how many of these will
an assigned bell be rung last?

Answer. This is the number of permutations of 8 distinct things taken 8 at a time $= \frac{8!}{(8-8)!} = 40,320$.
In all $40,320$ permutations, all 8 bells are equally
likely to be chosen. So for a given bell $\frac{40,320}{8} =$
$5,040$ of the permutations will have that bell rung
last.

Exercise 18. There are 3 teetotums having 6, 8, and
10 sides. In how many ways can they fall? And
of those ways, how many will have at least two
1's show up?

Answer. They can fall in $6 \cdot 8 \cdot 10 = 480$ ways. There
are 4 categories of ways to get at least two 1's.
The first category is to get three 1's, for which
there is only one way. The second category is
for the first 2 teetotums to have 1's and the last
anything but a 1, for which there are 9 ways. The
third category is for the first and third teetotum
to have 1's, and the second anything but a 1, for
which there are 7 ways. The fourth category is for
the second and third teetotums to have 1's, and

the first anything but a 1, for which there are 5 ways. So the total number of ways is $1+9+7+5 = 22$.

Exercise 19. Having cloth of 5 different colors, in how many ways can I choose 3 colors for a tricolor flag?

Answer. $\binom{5}{3} = \frac{5!}{3! \cdot 2!} = 10$ ways.

Exercise 20. In how many ways can I arrange vertically the 3 colors as 3 horizontal strips?

Answer. This is the number of permutations of 3 unique things taken 3 at a time $= \frac{3!}{(3-3)!} = 6$ ways. So the total number of different flags I can make is $10 \cdot 6 = 60$.

Exercise 21. I would like to publish a set of dictionaries to translate from one language to any other. If I limit myself to 5 languages, how many dictionaries do I need to publish?

Answer. This is the number of permutations of 5 unique things taken 2 at a time $= \frac{5!}{(5-2)!} = 20$ dictionaries.

Exercise 22. If I extend my limit to 10 languages, how many more dictionaries do I need to publish?

Answer. For 10 languages I need to publish $\frac{10!}{(10-2)!} = 90$ dictionaries, which is 70 more dictionaries than if I had only 5 languages.

Exercise 23. A father lives near 3 boys schools and 4 girls schools. In how many ways can he send his 3 sons and 2 daughters to school?

Answer. $3^3 \cdot 4^2 = 432$ ways.

Exercise 24. With 300 names to choose from, in how many ways can a child be named if we don't use more than 3 names at once?

Answer. Using only one name, there are 300 choices. Using 2 names there are $\frac{300!}{(300-2)!} = \frac{300!}{298!} = 89,700$ choices, and using 3 names there are $\frac{300!}{(300-3)!} = \frac{300!}{297!} = 26,730,600$ choices. So the total number of ways a child can be named is $26,730,600 + 89,700 + 300 = 26,820,600$.

Exercise 25. In how many arrangements can 4 people sit at a round table so that none will have the same neighbors?

Answer. If I am one of the 4 people, the other 3 people can be arranged in $3! = 6$ ways, but half of those ways are the same arrangements with respect to rotation which doesn't change the neighbors. So the number of arrangements is $\frac{3!}{2} = 3$.

Exercise 26. In how many ways can 7 people sit as described in the previous question? In how many of these ways will 2 assigned people be neighbors? In how many of these ways will an assigned person have the same 2 neighbors?

Answer. If I am one of the 7 people, the other 6 can be arranged in $6! = 720$ ways, but half of those have the same neighbors, so the number of arrangements is $\frac{720}{2} = 360$. Out of these 360 arrangements I have 720 neighbors, and if I am one of 2 assigned people, the other assigned person must be my neighbor $\frac{720}{6} = 120$ times. For the case of an assigned person having the same 2 neighbors, there are $4! = 24$ arrangements of the other 4 people, with 2 arrangements of the same neighbors, but half of all these arrangements have the same neighbors, so the result is $\frac{24 \cdot 2}{2} = 24$ ways.

Exercise 27. 5 women and 3 men get together to play croquet. In how many ways can they divide themselves into 2 groups of 4 so that the men are not all on the same side?

Answer. There is only one way to split the men into 2 groups so that they're not on the same side, 2 on one side, and one on the other. With 3 men, this can be done in $\binom{3}{2} = \frac{3!}{2!} = 3$ ways. The side with 2 men on it has room for 2 women, and the side with one man on it has room for 3 women, so the women can be split in $\binom{5}{2} = \frac{5!}{2! \cdot 3!} = 10$ ways. Thus the total number of ways is $3 \cdot 10 = 30$.

Exercise 28. I have 6 packages to be delivered with a choice of 3 carriers. In how many ways can I send the packages?

Answer. For each package, I have a choice of 3 carriers, so the total number of ways is $3^6 = 729$.

Exercise 29. Spike has 7 different books, while Spud has 9 different books. In how many ways can one of Spike's books be exchanged for one of Spud's?

Answer. Spike can choose one of his 7 books to exchange, while Spud can choose one of his 9, so there are $7 \cdot 9 = 63$ ways.

Exercise 30. For the previous question, how many ways can 2 of Spike's books be exchanged for 2 of Spud's?

Answer. Spike can choose 2 out of 7 of his books, which makes $\binom{7}{2} = \frac{7!}{2! \cdot 5!} = 21$ ways. Spud can choose 2 out of 9, which makes $\binom{9}{2} = \frac{9!}{2! \cdot 7!} = 36$ ways. The total number of ways is then $21 \cdot 36 = 756$.

Exercise 31. Five guys, Spike, Spud, Skeeter, Sparky, and Spivy will be speaking at a meeting. How many ways can they take their turn without Spud speaking before Spike?

Answer. Without restrictions, the total number of ways the men can speak is $\frac{5!}{(5-5)!} = 120$. In half of these ways Spike will speak before Spud, and vice versa for the other half. So there are $\frac{120}{2} = 60$ ways without Spud speaking before Spike.

Exercise 32. For the previous question, in how many ways can Spike speak immediately before Spud?

Answer. If we consider Spike and Spud to be one speaker, then there are $\frac{4!}{(4-4)!} = 24$ ways they can be arranged with the other 3 speakers. This is the number of ways Spike can speak immediately before Spud.

Exercise 33. In a company of soldiers there are 3 officers, 4 sergeants, and 60 privates. In how many ways can we form a detachment consisting of an officer, 2 sergeants, and 20 privates? In how many of those ways will the captain and senior sergeant appear?

Answer. Such a detachment can be formed in $\binom{3}{1} \cdot \binom{4}{2} \cdot \binom{60}{20} = 3 \cdot 6 \cdot 4,191,844,505,805,495 = 75,453,201,104,498,910$ ways. Of these ways, the captain appears once every 3 times, and the senior sergeant appears half the time. So the

number of ways the captain and senior sergeant appear is the total ways above divided by 6 or $\frac{75,453,201,104,498,910}{6} =$
$12,575,533,517,416,485.$

Exercise 34. From 12 ladies and 15 gents, how many ways can 4 ladies and 4 gents be chosen for a dance?

Answer. $\binom{12}{4} \cdot \binom{15}{4} = 495 \cdot 1,365 = 675,675$ ways.

Exercise 35. A guy belongs to a fruit-of-the-month club which has 30 fruits available to choose from. If he selects 5 fruits a month, forming a different group of 5 each month, for how many months can he do this?

Answer. $\binom{30}{5} = 142,506$ months, or 11,875.5 years.

Exercise 36. Show that the letters of *anticipation* can be arranged in 3 times as many ways as the letters of *commencement*.

Answer. *anticipation* has 12 letters, with 2 a's, 2 n's, 2 t's, 3 i's, and single occurrences of the other 3 letters. The letters of this word can be arranged in $\binom{12}{2,2,2,3} = 9,979,200$ ways. *commencement* has 12 letters, with 2 c's, 3 m's, 3 e's, 2 n's, and single occurrences of the other 2 letters. The letters of this word can be arranged in $\binom{12}{2,3,3,2}$ = $3,326,400$ ways, which is $\frac{1}{3}$ the number of ways for the previous word.

Exercise 37. How many 5 letter words can be made from 26 letters with repetitions allowed, but with no 2 identical letters as neighbors?

Answer. The first letter can be chosen in 26 ways, the second in 25 ways (omitting the letter just chosen), the third in 25 ways (omitting the letter just chosen), etc. So the number of such 5 letter words is $26 \cdot 25^4 = 10,156,250$.

Exercise 38. 8 people will row a boat, but 2 can only row on the left side, and three only on the right side, with the rest able to row on either side. In how many ways can the people be arranged?

Answer. Of the 3 remaining people who can row on either side, 2 of them are needed for the left side, and one on the right side. These 3 people can therefore be arranged in $\binom{3}{2} = 3$ ways. The 4 people on the left side can be arranged in $4! = 24$ ways, and the 4 people on the right side also in 24 ways. So the total number of possible arrangements is $3 \cdot 24 \cdot 24 = 1,728$.

Exercise 39. I have 3 copies of Tarzan, 2 copies of Robinson Crusoe, and 1 copy of Don Quixote. In how many ways can I give them to a class of 12 students, (1) so that no student gets more than 1 book and (2) so that no student receives more than 1 copy of any book?

Answer. (1) If no student gets more than 1 book, then there are $\binom{12}{3} \cdot \binom{9}{2} \cdot \binom{7}{1} = 220 \cdot 36 \cdot 7 = 55,440$ ways to give them. (2) If no student gets more than 1 copy of any book, then there are $\binom{12}{3} \cdot \binom{12}{2} \cdot \binom{12}{1} = 220 \cdot 66 \cdot 12 = 174,240$ ways.

Exercise 40. In how many ways can a set of 12 black checkers and 12 white checkers be placed on the 32 black squares of a checker board?

Answer. We divide the 32 black squares of the checker board into 3 different subsets, the first subset of size 12 contains black checkers, the second subset of size 12 contains white checkers, and the remaining 8 are empty. The number of combinations of these 3 subsets is therefore $\binom{32}{12,12,8} = 28,443,124,054,800$.

Exercise 41. In how many ways can the letters of the word *possessions* be arranged?

Answer. This word has 11 letters, with 2 *o*'s, 5 *s*'s, and single occurrences of the remaining letters. So the letters of this word can be arranged in $\binom{11}{2,5} = 166,320$ ways.

Exercise 42. In how many ways can the letters of the word *cockatoo* be arranged?

Answer. This word has 8 letters, with 2 *c*'s, 3 *o*'s, and single occurrences of the remaining letters. So the letters of this word can be arranged in $\binom{8}{2,3} = 3,360$ ways.

Exercise 43. Show that the letters of *cocoon* can be arranged in twice as many ways as the letters of *cocoa*.

Answer. *cocoon* has 6 letters, with 2 *c*'s, 3 *o*'s, and one *n*, which allows $\binom{6}{2,3} = 60$ arrangements. *cocoa* has 5 letters, with 2 *c*'s, 2 *o*'s, and one *a*, which allows $\binom{5}{2,2} = 30$ arrangements, and that's half the arrangements of the word *cocoon*.

Exercise 44. In how many ways can you arrange the letters of *pallmall* without letting all the *l*'s come together?

Answer. *pallmall* has 8 letters with 2 *a*'s, 4 *l*'s, and single occurrences of the remaining 2 letters. Without restrictions, there are $\binom{8}{2,4} = 840$ arrangements. But there are $\binom{5}{2} = 60$ arrangements with all *l*'s together, so the number of arrangements without all the *l*'s together is $840 - 60 = 780$.

Exercise 45. In how many ways can you arrange the letters
oiseau so that the vowels are in alphabetical order?

Answer. This word contains 5 vowels and 1 consonant. If we put the vowels in alphabetical order *aeiou* then the consonant *s* is the only other letter to be arranged. It can be placed between these vowels or on the ends, or 6 ways.

Exercise 46. In how many ways can you arrange the letters of *cocoa* so that *a* is in the middle?

Answer. With *a* in the middle, there are 2 letters on either side, that are selected from a pair of *c*'s and a pair of *o*'s. This selection can be done in $\binom{4}{2,2} = 6$ ways.

Exercise 47. In how many ways can you arrange the letters of *quartus* so that *q* is followed by *u*?

Answer. Keeping *qu* together, we can consider it one letter, so there are 6 letters instead of 7. The number of arrangements then are $\frac{6!}{(6-6)!} = \frac{6!}{0!} = 720$.

Exercise 48. In how many ways can you arrange the letters of *ubiquitous* so that *q* is followed by *u*?

Answer. Keeping *qu* together, we can consider it one letter, so there are 9 letters instead of 10. Of these 9 letters, there are 2 *u*'s, 2 *i*'s, and the rest occur only once. The number of arrangements then are $\binom{9}{2,2} = 90,720$.

Exercise 49. In how many ways can you arrange the letters of *quisquis* so that each *q* is followed by *u*?

Answer. Keeping both *qu* pairs together, we can consider each pair to be a letter, so instead of 8 letters, there are 6. Of these 6 letters, there are 2 *qu*'s, 2 *i*'s, and 2 *s*'s. The number of arrangements then are $\binom{6}{2,2,2} = 90$.

Exercise 50. In how many ways can you arrange the letters of *indivisibility* without letting 2 *i*'s be together?

Answer. This word contains 14 letters, 6 of which are *i*'s, the remaining 8 occurring only once. The remaining 8 letters can be arranged in $8! = 40,320$ ways. For each of these arrangements the 6 *i*'s can be placed in between the 8 letters or on either end, for 9 possible positions. The number of ways this can be done is $\binom{9}{6} = 84$. So the total number of ways is $40,320 \cdot 84 = 3,386,880$.

Exercise 51. In how many ways can you arrange the letters of *facetious* without letting 2 vowels be together?

Answer. *facetious* contains 9 letters, with 5 vowels and 4 consonants. The 4 consonants can be arranged in $4! = 24$ ways. The 5 vowels, which are all unique, can be placed in between the 4 consonants or on the ends, making 5 possible positions. This can be done in $\frac{5!}{(5-5)!} = \frac{5!}{0!} = 120$ ways. The total number of ways is then $24 \cdot 120 = 2,880$.

Exercise 52. In how many ways can you rearrange the letters of *facetious* without changing the order of the vowels?

Answer. The 9 letters of *facetious*, which are all unique, can be arranged in $\frac{9!}{(9-9)!} = 9!$ ways. The 5 vowels can be arranged in $\frac{5!}{(5-5)!} = 5!$ ways. The number of arrangements of the 9 letters, with the vowels in their original order, is then $\frac{9!}{5!} = 3,024$. One of these arrangements is just the original word, so the number of rearrangements is 3,023.

Exercise 53. In how many ways can you rearrange the letters of *abstemiously* without changing the order of the vowels?

Answer. The 12 letters of *abstemiously*, which are all unique except for 2 *s*'s, can be arranged in

$\binom{12}{2,1,1,\dots,1} = 239{,}500{,}800$ ways. The 5 vowels of this word can be arranged in $\frac{5!}{(5-5)!} = 5!$ ways. The number of arrangements of the 12 letters, with the vowels in their original order, is then $\frac{239{,}500{,}800}{5!} = 1{,}995{,}840$. One of these arrangements is just the original word, so the number of rearrangements is 1,995,839.

Exercise 54. In how many ways can you rearrange the letters of *parallelism* without changing the order of the vowels?

Answer. The 11 letters of *parallelism*, which are all unique except for 2 *a*'s and 3 *l*'s, can be arranged in $\binom{11}{2,3} = 3{,}326{,}400$ ways. The 4 vowels of this word can be arranged in $\binom{4}{2,1,1} = 12$ ways. The number of arrangements of the 11 letters, with the vowels in their original order, is then $\frac{3{,}326{,}400}{12} = 277{,}200$. One of these arrangements is just the original word, so the number of rearrangements is 277,199.

Exercise 55. In how many ways can you rearrange the letters of *almost*, maintaining the current separation of the vowels?

Answer. *almost* has 6 letters, with 4 consonants and 2 vowels. The 4 consonants can be arranged in $\frac{4!}{(4-4)!} = 4! = 24$ ways. The 2 vowels, being in the first and fourth positions, can be arranged in 2 ways, but they can also occupy the second and fifth positions, as well as the third and sixth positions, so the total number of arrangements are $24 \cdot 2 \cdot 3 = 144$ ways. One of these ways is the original, so there are 143 ways to rearrange.

Exercise 56. In how many ways can you rearrange the letters of *logarithms*, so that consonants occupy the second, fourth, and sixth positions (where the vowels currently are)?

Answer. *logarithms* has 10 letters, with 7 consonants and 3 vowels. The 3 positions where the consonants must be, can be filled in $\frac{7!}{(7-3)!} = \frac{7!}{4!} = 210$ ways. The remaining 7 positions can be filled with any of the remaining 7 letters, for $\frac{7!}{(7-7)!} = \frac{7!}{0!} = 5,040$ ways. So the total arrangements are $210 \cdot 5,040 = 1,058,400$ ways.

Exercise 57. In how many ways can 2 consonants and a vowel be chosen from the word *logarithms*, and how many of those ways has the letter *s* in it?

Answer. 2 consonants can be chosen from the 7 in $\binom{7}{2} = \frac{7!}{2! \cdot 5!} = 21$ ways. A vowel can be chosen from the 3 in $\binom{3}{1} = \frac{3!}{2!} = 3$ ways. So the 2 consonants and a vowel can be chosen in $21 \cdot 3 = 63$ ways. If *s* is always chosen as one of the consonants, then there are 6 choices for the second consonant, and 3 choices for the vowel, for a total of $6 \cdot 3 = 18$ ways.

Exercise 58. In how many ways can you arrange the letters of *syzygy* without all 3 y's being bunched together?

Answer. Without the restriction, there are $\binom{6}{3,1,1,1} = \frac{6!}{3!} = 120$ ways. The number of ways all 3 y's can be bunched together is $\frac{4!}{(4-4)!} = \frac{4!}{0!} = 24$. So the number of arrangements without all 3 y's bunched together is $120 - 24 = 96$ ways.

Exercise 59. For the previous question, what is the number of ways without any 2 y's being bunched together?

Answer. The 3 letters s, z, and g can be ordered in $\frac{3!}{(3-3)!} = 6$ ways. The 3 y's can be placed in any of 4 positions around them, which can be done in $\binom{4}{3} = 4$ ways. So the total number of arrangements is $6 \cdot 4 = 24$.

Exercise 60. In how many ways can you arrange the letters of the words *choice and chance* without any 2 c's being bunched together?

Answer. There are 15 letters, with 4 c's, 2 h's, 2 e's, 2 a's, 2 n's, and single occurrences for the remaining letters. Without the c's, there are 11 letters that can be arranged in $\binom{11}{2,2,2,2} = \frac{11!}{2^4} = 2,494,800$ ways. Between these 11 letters and on the ends, there are 12 positions we can place the 4 c's, with $\binom{12}{4} = 495$ ways to make these placements. The total number of ways to make the arrangements is then $2,494,800 \cdot 495 = 1,234,926,000$.

Exercise 61. How many ways can you bundle 10 books into 5 parcels of 2 books each?

Answer. $\binom{10}{2,2,2,2,2} = \frac{3,628,800}{2^5} = 113,400$ ways. In this case the order of the 5 parcels doesn't matter, so we need to divide by 5!, which gives $\frac{113,400}{120} = 945$ ways.

Exercise 62. How many ways can you bundle 9 books into 4 parcels of 2 books each, with one left over?

Answer. $\binom{9}{2,2,2,2,1} = \frac{362,880}{2^4} = 22,680$ ways. The order of the 4 parcels with 2 books each doesn't matter, so we need to divide by 4!, which gives $\frac{22,680}{24} = 945$ ways.

Exercise 63. How many ways can you bundle 9 books into 3 parcels of 3 books each?

Answer. $\binom{9}{3,3,3} = \frac{362,880}{6^3} = 1,680$ ways. The order of the 3 parcels doesn't matter, so we need to divide by 3!, which gives $\frac{1,680}{6} = 280$ ways.

Exercise 64. How many ways can 10 husband and
    wife couples be formed into 5 groups with 2 men
    and 2 women in each group?

Answer. This problem can be broken down into the
    number of ways of taking 10 distinct men and
    forming 5 distinct groups of 2 men each, and sim-
    ilarly for the women. For the men, this can be
    done in $\binom{10}{2,2,2,2,2} = \frac{10!}{2^5} = 113,400$ ways. The or-
    der of the 5 groups doesn't matter, so we must
    divide by 5!, giving $\frac{113,400}{120} = 945$ ways. The same
    result is gotten for the women in the same way.
    The number of ways to pair up the 5 groups of
    men with the 5 groups of women is 5!, so the total
    number of ways is $945 \cdot 945 \cdot 120 = 107,163,000$.

Exercise 65. In how many of these ways will a given
    man find himself in the same group as his wife?

Answer. A given man has 2 women in his group. Since
    the choice of the 2 women out of 10 is unbiased,
    he will on average find himself in the same group
    as his wife $\frac{1}{5}$ of the time. So the number of ways
    is $\frac{107,163,000}{5} = 21,432,600$.

Exercise 66. In how many of the ways will 2 given
    men find themselves in the same group as their
    wives?

Answer. If 2 given men and their wives make up one group, then the only variations are the number of ways that 8 men and 8 women can form 4 groups of 2 men and 2 women. This can be done in $\frac{\left(\binom{8}{2,2,2,2}\right)^2}{(4!)^2} \cdot 4! = 264,600$ ways (see the question before last for detailed explanation).

Exercise 67. In how many ways can you select 6 bandanas, with repetitions allowed, at a shop that has 7 types available?

Answer. This question is equivalent to "How many multisets of size 6 can you create by sampling with replacement from a set of size 7?". The answer is $\binom{6+7-1}{7-1} = \binom{12}{6} = \frac{12!}{6! \cdot 6!} = 924$ ways.

Exercise 68. From a 10 member choir, in how many ways can you select a different group of 6 every day for 3 days?

Answer. The number of subsets of size 6 from a set of size 10 is $\binom{10}{6} = \frac{10!}{6! \cdot 4!} = 210$. Selecting 3 items from 210 can be done in $\frac{210!}{(210-3)!} = \frac{210!}{207!} = 9,129,120$ ways.

Exercise 69. A man has 6 friends, and invites 3 of them to dinner every day for 20 days. How many ways can he do this while selecting a different group of 3 each time?

Answer. The number of subsets of size 3 from a set of size 6 is $\binom{6}{3} = \frac{6!}{3! \cdot 3!} = 20$. Selecting 20 items from 20 can be done in $\frac{20!}{(20-20)!} = \frac{20!}{0!} = 2,432,902,008,176,640,000 \approx 2.43 \times 10^{18}$ ways.

Exercise 70. What is the total number of selections you can make from the letters *ned needs nineteen nets*?

Answer. There are 20 letters, with 6 *n*'s, 7 *e*'s, 2 *d*'s, 2 *s*'s, 1 *i*, and 2 *t*'s. From the *n*'s we can choose none or up to 6, making 7 choices. For the *e*'s we can choose none or up to 7, making 8 choices, and similarly for the remaining letters. So the total number of selections we can make is $7 \cdot 8 \cdot 3 \cdot 3 \cdot 2 \cdot 3 = 3,024$. One of these selections includes the case of not choosing any letters at all, so the total is 3,023 selections.

Exercise 71. What is the total number of selections you can make from the letters *daddy did a nice deed*?

Answer. There are 17 letters, with 7 *d*'s, 2 *a*'s, 1 *y*, 2 *i*'s, 1 *n*, 1 *c*, and 3 *e*'s. Similarly to the last question, the total number of selections we can make is $8 \cdot 3 \cdot 2 \cdot 3 \cdot 2 \cdot 2 \cdot 4 = 2,304$. One of these selections includes the case of not choosing any letters at all, so the total is 2,303 selections.

Exercise 72. From the letters in the last question, how many selections of 3 letters can be made?

Answer. The 3 chosen letters can fall into any of 3 categories:

1. all 3 letters identical
2. 2 letters identical, one different
3. all 3 letters different

For the first category there are 2 ways. For the second category there are 4 ways to choose 2 identical letters, and 6 ways to choose the different letter, which makes $4 \cdot 6 = 24$ ways. For the third category there are $\binom{7}{3} = \frac{7!}{3! \cdot 4!} = 35$ ways to choose 3 different letters. The total number of selections that can be made is then $2 + 24 + 35 = 61$.

Exercise 73. From the last question, how many arrangements of 3 letters can be made?

Answer. For category (1) there are 2 arrangements of the 2 ways. For category (2) there are $24 \cdot \binom{3}{2,1} = 72$ arrangements of the 24 ways. And for category (3) there are $35 \cdot 3! = 210$ arrangements of the 35 ways. The total number of arrangements is then $2 + 72 + 210 = 284$.

Exercise 74. Show that there are 8 subsets of size 3, from the letters of *veneer*, and 16 subsets of size 4 from the letters of *veneered*.

Answer. *veneer* has 6 letters with 3 *e*'s, and the rest single occurences. Subsets of size 3 fall into 3 categories:

1. All 3 letters are the same. There is only one way for this to occur.

2. Just 2 letters are the same, one is different. There are 3 ways for this to happen.

3. All 3 letters are different. There are $\binom{4}{3} = \frac{4!}{3!} = 4$ ways for this.

The total number of subsets of size 3 is then $1 + 3 + 4 = 8$. *veneered* has 8 letters with 4 *e*'s, and the rest single occurences. There are 4 categories for subsets of size 4:

1. All 4 letters are the same. There is only one way for this to occur.

2. Just 3 letters are the same, one is different. There are $\binom{4}{1} = 4$ ways for this to happen.

3. Just 2 letters are the same, 2 are different. There are $\binom{4}{2} = 6$ ways for this to happen.

4. All 4 letters are different. There are $\binom{5}{4} = \frac{5!}{4!} = 5$ ways for this to happen.

The total number of subsets of size 4 is then $1 + 4 + 6 + 5 = 16$.

Exercise 75. How many subsets of size 3 are there in the letters *wedded*?

Answer. *wedded* has 6 letters with 2 *e*'s, 3 *d*'s, and one *w*. Subsets of size 3 fall into 3 categories:

1. All 3 letters are the same. There is only one way for this to occur.

2. Just 2 letters are the same, one is different. There are $2 \cdot 2 = 4$ ways for this to occur.

3. All 3 letters are different. There is $\binom{3}{3} = 1$ way for this to occur.

The total number of subsets of size 3 is then $1 + 4 + 1 = 6$.

Exercise 76. How many subsets of size 4 are there in the letters *redeemed*?

Answer. *redeemed* has 8 letters with 4 *e*'s, 2 *d*'s, and one occurrence each of *r* and *m*. Subsets of size 4 fall into 5 categories:

1. All 4 letters are the same. There is only one way for this to occur.

2. Just 3 letters are the same, one is different. There are 3 ways for this to occur.

3. Only 2 letters are the same, 2 are different. There are $2 \cdot \binom{3}{2} = 6$ ways for this to occur.

4. 2 letters are the same, the other 2 are the same. There is one way for this to occur.

5. All 4 letters are different. There is $\binom{4}{4} = 1$ way for this to occur.

The total number of subsets of size 4 is then $1 + 3 + 6 + 1 + 1 = 12$.

Exercise 77. How many subsets of size 5 are there in the letters *ever esteemed*?

Answer. *ever esteemed* has 12 letters with 6 *e*'s, and the rest single occurrences. Subsets of size 5 fall into 5 categories:

1. All 5 letters are the same. There is only one way for this to occur.
2. Just 4 letters are the same, one is different. There are 6 ways for this to occur.
3. Only 3 letters are the same, 2 are different. There are $\binom{6}{2} = 15$ ways for this to occur.
4. 2 letters are the same, the other 3 are different. There are $\binom{6}{3} = 20$ ways for this to occur.
5. All 5 letters are different. There are $\binom{7}{5} = 21$ ways for this to occur.

The total number of subsets of size 5 is then $1 + 6 + 15 + 20 + 21 = 63$.

Exercise 78. How many ways can 3 people distribute the letters of *ever esteemed* among each other?

Answer. *ever esteemed* has 12 letters with 6 e's, and the rest single occurrences. Distributing the 6 e's is equivalent to the question "How many ways can you place 6 identical balls into 3 distinct bins with no restrictions?". The answer is $\binom{6+3-1}{3-1} = \binom{8}{2} = 28$ ways. Distributing the remaining 6 unique letters is equivalent to the question "How many ways can you place 6 distinct balls into 3 distinct bins with no restrictions?". The answer is $3^6 = 729$ ways. The total number of ways is then $28 \cdot 729 = 20,412$. But this includes the cases where either one or 2 people end up with nothing. There are 3 ways for 2 people to end up with nothing. For only one person to end up with nothing, the other 2 people must get something, and that can be done in $\binom{6+2-1}{2-1} \cdot 2^6 - 2 = 7 \cdot 64 - 2 = 446$ ways, where the 2 is subtracted for the cases of either of the 2 people getting everything. Since there are 3 people who can end up with nothing, we must multiply this result by 3 to get $446 \cdot 3 = 1,338$. So removing the cases where someone ends up with nothing, we get $20,412 - 3 - 1,338 = 19,071$ ways.

Exercise 79. For the previous question, how many ways can the letters be divided so that each person gets 4?

Answer. This is equivalent to the question "If you have 12 balls, with 6 being distinct and 6 identical, how many ways can you put 4 balls each into 3 distinct urns?". We'll look at this problem from the standpoint of putting the distinct balls into the urns first, then topping off with identical balls to end with 4 balls in each urn. If there were no restrictions on the number of distinct balls placed into each urn, then the answer would be $3^6 = 729$. Now we just need to subtract the number of ways that any urn can get 6 or 5 balls. All 6 balls can end up in a certain urn in 1 way. 5 balls can end up in a certain urn in $\binom{6}{5} = 6$ ways, with 2 ways to place the remaining ball, making $6 \cdot 2 = 12$ ways. Because we have 3 urns, the number of ways to get 6 or 5 balls in an urn is then $3 \cdot (1 + 12) = 39$ ways. The total number of ways is then $729 - 39 = 690$. Filling each urn up to the required 4 with identical balls can only be done in one way, so the result is still 690.

Exercise 80. Albert has the six letters *esteem*, Bob has the six letters *feeble*, and Carter has the six letters *veneer*. In how many ways can the letters be redistributed so each of them still has six letters?

Answer. There are a total of 18 letters with 9 of them being *e*'s and the other 9 all different. First distribute the different letters so that no one has more than 6 and then distribute the *e*'s to give everyone exactly 6 letters. The 9 different letters can be distributed in a total of $3^9 = 19,683$ ways but in some of these ways a person gets more than 6 letters. Albert can get 9 letters in only one way. He can get 8 letters in 9 ways with Bob and Carter getting the remaining letter in 2 ways. He can get 7 letters in $\binom{9}{7} = 36$ ways and the other two get the remaining 2 letters in $2^2 = 4$ ways. The number of ways Albert gets more than 6 letters is then $1 + 9 \cdot 2 + 36 \cdot 4 = 163$. The same applies to Bob and Carter. This means there are $3 \cdot 163 = 489$ ways in which someone gets more than 6 of the different letters. The different letters can therefor be distributed in $19,683 - 489 = 19,194$ correct ways. For each of these ways there is only one way to distribute the e's so that each person has exactly 6 letters. The answer to the question is then $19,194$.

Exercise 81. How many ways can you arrange the letters in the word *falsity* so that the consonants f, l, s, t and the vowels a, i, y keep the same order?

Answer. You can choose 4 of the 7 letters to be consonants and the remaining 3 to be vowels in $\binom{7}{4} = 35$ ways. For each of these ways there is only one way to place the consonants and vowels to keep the same order so the answer is 35.

Exercise 82. How many ways can you arrange the letters in the word *affection* so that the vowels keep their order and the two *f*'s are seperated?

Answer. There are a total of 9 letters with 5 consonants and 4 vowels. The number of ways places for the consonants can be selected is $\binom{9}{5}$. Assuming the 2 *f*'s are distinguishable, there are 5! ways to arrange the consonants once the places have been chosen. Distinguishability assumption for the *f*'s can be removed by dividing by 2. For every way of arranging the consonants there is only one way to arrange the vowels to keep their order. So the number of ways to arrange the letters without restriction on keeping the *f*'s apart is $\binom{9}{5}\frac{5!}{2} = \frac{9!}{4!2} = 7,560$. To find the number of these arrangements where the *f*'s are together, treat *ff* as a new consonant. Now there are 8 letters with 4 consonants and 4 vowels. The number of ways they can be arranged is $\binom{8}{4}4! = \frac{8!}{4!} = 1,680$. Subtract this from the previous number to get the number of arrangements with the *f*'s seperated: $7,560 - 1,680 = 5,880$.

Exercise 83. How many ways can you arrange the letters of *kaffeekanne* so that each arrangement alternates between consonant and vowel?

Answer. There are 6 consonants *kkffnn* that can be arranged in $\binom{6}{2,2,2} = \frac{6!}{2^3} = 90$ ways. There are 5 vowels *aaeee* that can be arranged in $\binom{5}{2,3} = \frac{5!}{12} =$ 10 ways. For each arrangement of consonants, pick an arrangement of vowels to place between the consonants. The number of ways to do this is then $90 \cdot 10 = 900$.

Exercise 84. How many arrangements of the word *delete* keep the order of the consonants?

Answer. *delete* has 6 letters, with 3 consonants *dlt*, and 3 vowels which are all *e*'s. This problem can be thought of in terms of 6 urns, with 3 of the urns to be chosen to place the consonants in their fixed order. 3 urns out of 6 can be chosen in $\binom{6}{3} = 20$ ways. The remaining 3 urns are filled each with one *e*, which can be done in only one way. So the total number of arrangements is 20.

Exercise 85. For the previous question, how many arrangements are there if we add the restriction of not having any 2 *e*'s together?

Answer. There are 4 ways: *delete*, *edelet*, *edlete*, and *edelte*.

Exercise 86. How many ways can the letters of *delirious* be arranged keeping the vowels and consonants in their original order?

Answer. *delirious* has 9 letters with 4 consonants and 5 vowels. Places for the consonants can be chosen in $\binom{9}{4} = 126$ ways. Vowels and consonants must be kept in order so there is only one way to arrange them for each of these ways. The answer is then 126.

Exercise 87. For the previous question, what is the number of ways if we exclude arrangements with 2 *i*'s side by side?

Answer. Treat the 2 *i*'s as a single vowel, so there are 8 letters with 4 consonants and 4 vowels. The consonant places can be chosen in $\binom{8}{4} = 70$ ways. In each of these ways the consonants and vowels can only be filled in one way. So there are 70 arrangements where the 2 *i*'s appear together. Subtract this from the previous answer to get $126 - 70 = 56$ ways to arrange things without the 2 *i*'s together.

Exercise 88. How many ways can the letters *fulfil* be arranged so that no 2 consecutive letters are the same?

Answer. *fulfil* has 6 letters, with 2 *f*'s, 1 *u*, 2 *l*'s, and 1 *i*. Without restriction, the letters can be arranged in $\binom{6}{2,2} = \frac{6!}{4} = 180$ ways. Treating *ff* as a single letter, the number of arrangements are $\binom{5}{2,1,1,1} = \frac{5!}{2} = 60$. Likewise, treating ll as a single letter, the number of arrangements are $\binom{5}{2,1,1,1} = \frac{5!}{2} = 60$. Treating ff and ll as single letters, the number of arrangements are $\binom{4}{1,1,1,1} = 4! = 24$. The number of arrangements where ff, ll, or both appear is then $60 + 60 - 24 = 96$. So the number of arrangements where neither ff or ll appear is $180 - 96 = 84$.

Exercise 89. How many ways can the letters $murmur$ be arranged so that no 2 consecutive letters are alike?

Answer. $murmur$ has 6 letters, with 2 m's, 2 u's, and 2 r's. The number of unrestricted arrangements is $\binom{6}{2,2,2} = \frac{6!}{8} = 90$. Let

A = set of arrangements with 2 consec. m's

B = set of arrangements with 2 consec. u's

C = set of arrangements with 2 consec. r's

The number of members of A, B, and C is then

$$|A| = |B| = |C| = \binom{5}{2,2,1} = \frac{120}{4} = 30$$

The number of common members between the

sets is

$$|A \cap B| = |A \cap C| = |B \cap C|$$
$$= \binom{4}{2} = 12$$
$$|A \cap B \cap C| = 3! = 6$$

$A \cup B \cup C =$
set of arrangements with at least 1 double letter

$$|A \cup B \cup C| = |A| + |B| + |C|$$
$$- |A \cap B| - |A \cap C|$$
$$- |B \cap C|$$
$$+ |A \cap B \cap C|$$
$$= 3 \cdot 30 - 3 \cdot 12 + 6$$
$$= 3 \cdot 18 + 6$$
$$= 54 + 6 = 60$$

Subtracting this from the unrestricted arrangements gives $90 - 60 = 30$.

Exercise 90. How many ways can you select 4 letters from *murmur* and how many arrangements are possible?

Answer. There are 3 letters with multiplicities of 2 each. When selecting 4 letters from these 6, there will be either 2 letters with multiplicity 2 each, or one with multiplicity 2 and the other 2 different. 2 with multiplicity 2 can be selected in $\binom{3}{2} = 3$ ways. One with multiplicity 2, and the other 2 different can be selected in $\binom{3}{1}\binom{2}{2} = 3$ ways. The number of selections is then $3 + 3 = 6$. There are $\binom{4}{2,2} = 6$ ways to arrange the $(2, 2)$ multiplicities, and $\binom{4}{2,1,1} = 12$ ways to arrange the $(2, 1, 1)$ multiplicities, so the number of arrangements is $3 \cdot 6 + 3 \cdot 12 = 18 + 36 = 54$.

Exercise 91. How many 4 letter words can you make from the letters $fulfil$?

Answer. $fulfil$ has 4 unique letters (f, l, i, u) with multiplicities $(2, 2, 1, 1)$ for a total of 6 letters. A selection of 4 letters can be made with multiplicities $(2, 2)$ or $(2, 1, 1)$, or $(1, 1, 1, 1)$. The $(2, 2)$ selection can be made in one way and the resulting letters can be arranged in $\binom{4}{2,2} = \frac{4!}{2^2} = 6$ ways. The $(2, 1, 1)$ selection can be made in $\binom{2}{1}\binom{3}{2} = 2 \cdot 3 = 6$ ways and the resulting letters arranged in $\binom{4}{2} = 12$ ways. The $(1, 1, 1, 1)$ selection can be made in one way, and the resulting letters arranged in $4! = 24$ ways. The total number of words is then $1 \cdot 6 + 6 \cdot 12 + 1 \cdot 24 = 6 + 72 + 24 = 102$.

Exercise 92. How many 5 letter words can you make from the letters $pallmall$?

Answer. There are 4 unique letters (l, a, m, p) with multiplicities $(4, 2, 1, 1)$ for a total of 8 letters. The multiplicities of a selection of 5 letters is shown in table 3.1. The total number of 5 let-

Mult	Selections	Arrangements	Select·Arrang
(4,1)	3	$\frac{5!}{4!} = 5$	15
(3,2)	1	$\frac{5!}{3! \cdot 2!} = 10$	10
(3,1,1)	3	$\frac{5!}{3!} = 20$	60
(2,2,1)	2	$\frac{5!}{2! \cdot 2!} = 30$	60
(2,1,1,1)	2	$\frac{5!}{2!} = 60$	120

Table 3.1: Multiplicities for Exercise 92.

ter words is then $15 + 10 + 60 + 60 + 120 = 265$.

Exercise 93. How many 4 letter words can you make from the letters $kaffeekanne$ excluding words with 3 e's in a row?

Answer. There are 5 unique letters (e, k, a, f, n) with multiplicities $(3, 2, 2, 2, 2)$ for a total of 11 letters. The multiplicities for a selection of 4 letters are shown in table 3.2. Without restrictions the number of words is $16 + 60 + 360 + 120 = 556$. There are 4 ways to select the 3 e's and one of the other letters. In each of these ways treat the 3 e's as a single letter then it can be arranged in 2 ways with the other letter. The number of words

Mult	Selections	Arrangements	Select·Arrang
(3,1)	4	$\frac{4!}{3!} = 4$	16
(2,2)	$\binom{5}{2} = 10$	$\frac{4!}{2! \cdot 2!} = 6$	60
(2,1,1)	$\binom{5}{1}\binom{4}{2} = 30$	$\frac{4!}{2!} = 12$	360
(1,1,1,1)	$\binom{5}{4} = 5$	$4! = 24$	120

Table 3.2: Multiplicities for Exercise 93.

with 3 e's together is then $42 = 8$. Subtracting this from the unrestricted word count gives $556 - 8 = 548$ words.

Exercise 94. How many 6 letter words can you make from the letters *nineteen tennis nets*?

Answer. There are 5 unique letters (n, e, t, i, s) with multiplicities $(6, 5, 3, 2, 2)$ for a total of 18 letters. When selecting 6 letters, the possible multiplicities, ways to select, and arrangements are in table 3.3. The total number of 6 letter words is then gotten by summing up the right column of the table, which is 12,289.

Exercise 95. How many 6 letter words can you make with the letters *littlepipe* if no letter can follow itself?

Answer. There are 5 unique letters (l, i, t, e, p) with multiplicities $(2, 2, 2, 2, 2)$ for a total of 10 letters.

Mult	Ways to Select	Arrangements	Select·Arrang
(6)	1	1	1
(5,1)	$\binom{2}{1}\binom{4}{1}=8$	$\frac{6!}{5!}=6$	48
(4,2)	$\binom{2}{1}\binom{4}{1}=8$	$\frac{6!}{4!\cdot2!}=15$	120
(4,1,1)	$\binom{2}{1}\binom{4}{2}=12$	$\frac{6!}{4!}=30$	360
(3,3)	$\binom{3}{2}=3$	$\frac{6!}{3!\cdot3!}=20$	60
(3,2,1)	$\binom{3}{1}\binom{4}{1}\binom{3}{1}=36$	$\frac{6!}{3!\cdot2!}=60$	2160
(3,1,1,1)	$\binom{3}{1}\binom{4}{3}=12$	$\frac{6!}{3!}=120$	1440
(2,2,2)	$\binom{5}{3}=10$	$\frac{6!}{(2!)^3}=90$	900
(2,2,1,1)	$\binom{5}{2}\binom{3}{2}=30$	$\frac{6!}{(2!)^2}=180$	5400
(2,1,1,1,1)	$\binom{5}{1}\binom{4}{4}=5$	$\frac{6!}{2!}=360$	1800

Table 3.3: Multiplicities for Exercise 94.

First find the number of 6 letter words with no restrictions. The possible multiplicities, ways to select, and arrangements are in table 3.4. The

Mult	Ways to Select	Arrangements	Select·Arrang
(2,2,2)	$\binom{5}{3}=10$	$\frac{6!}{(2!)^3}=90$	900
(2,2,1,1)	$\binom{5}{2}\binom{3}{2}=30$	$\frac{6!}{(2!)^2}=180$	5400
(2,1,1,1,1)	$\binom{5}{1}=5$	$\frac{6!}{2!}=360$	1800

Table 3.4: Multiplicities for Exercise 95.

number of unrestricted words is then gotten by summing up the right column of table 3.4, giving 8,100. With the $(2,2,2)$ multiplicity, we have 2 copies each of 3 unique letters. Let A, B, and

C be the sets containing words with the first, second, and third of the unique letters appearing double. We want to find

$$
\begin{aligned}
|A \cup B \cup C| = & |A| + |B| + |C| \\
& - |A \cap B| - |A \cap C| \\
& - |B \cap C| \\
& + |A \cap B \cap C|
\end{aligned}
$$

where

$$
|A| = |B| = |C| = \frac{5!}{2! \cdot 2!} = 30
$$
$$
|A \cap B| = |A \cap C| = |B \cap C|
$$
$$
= \frac{4!}{2!} = 12
$$
$$
|A \cap B \cap C| = 3! = 6
$$

so that

$$
|A \cup B \cup C| = 3 \cdot 30 - 3 \cdot 12 + 6 = 60
$$

There are 10 ways to have a $(2, 2, 2)$ multiplicity so there are a total of $60 \cdot 10 = 600$ restricted words for this multiplicity. With the $(2, 2, 1, 1)$ multiplicity there are 2 copies of 2 of the letters. Let A and B be the sets containing words where the first and second of these letters appear double. We want to find

$$
|A \cup B| = |A| + |B| - |A \cap B|
$$

where

$$|A| = |B| = \frac{5!}{2!} = 60$$
$$|A \cap B| = 4! = 24$$

so that

$$|A \cup B| = 2 \cdot 60 - 24 = 96$$

There are 30 ways to have a $(2, 2, 1, 1)$ multiplicity so there are $96 \cdot 30 = 2,880$ restricted words. For the $(2, 1, 1, 1, 1)$ multiplicity there are $5! = 120$ words with a double letter. There are 5 ways to get the multiplicity, so there are a total of $120 \cdot 5 = 600$ restricted words. The total number of restricted words is $600 + 2,880 + 600 = 4,080$. The number of acceptable words is then $8,100 - 4,080 = 4,020$.

Exercise 96. How many 5 letter words can you make with the letters *murmurer* so that the 3 *r*'s do not appear together?

Answer. There are 4 unique letters (r, m, u, e) with multiplicities $(3, 2, 2, 1)$ for a total of 8 letters. First find the number of 5 letter words with no restrictions. The possible letter multiplicities, ways to select and arrangements are in table 3.5. With no restrictions, the number of 5 letter words

Mult	Ways	Arrangements	Select·Arrang
(3,2)	2	$\frac{5!}{3!\cdot 2!} = 10$	20
(3,1,1)	3	$\frac{5!}{3!} = 20$	60
(2,2,1)	$\binom{3}{2}\binom{2}{1} = 6$	$\frac{5!}{(2!)^2} = 30$	180
(2,1,1,1)	$\binom{3}{1}\binom{3}{3} = 3$	$\frac{5!}{2!} = 60$	180

Table 3.5: Multiplicities for Exercise 96.

is the sum of the right column, which is 440. The $(3, 2)$ multiplicity will have 3 r's. Treat them as a single letter so that the number of words is $\binom{3}{2} = \frac{3!}{2!} = 3$. There are 2 ways to get this multiplicity, so the number of restricted words is $3 \cdot 2 = 6$. The $(3, 1, 1)$ multiplicity also has 3 r's. Treating them as a single letter, the number of words is $3! = 6$. There are 3 ways to get the multiplicity, so the number of restricted words is $6 \cdot 3 = 18$. The total number of restricted words is $6 + 18 = 24$. The number of acceptable words is then $440 - 24 = 416$.

Exercise 97. How many ways can you arrange the letters *quisquis* so that no letter follows itself?

Answer. There are 2 copies each of the letters (q, u, i, s). The number of unrestricted arrangements is

$$\frac{8!}{(2!)^4} = 2,520$$

Define
S_1 = number of words where 1 particular letter appears doubled.
S_2 = number of words where 2 particular letters appear doubled. Likewise for S_3 and S_4. The number of restricted words is then

$$\binom{4}{1}S_1 - \binom{4}{2}S_2 + \binom{4}{3}S_3 - \binom{4}{4}S_4$$

$$= 4S_1 - 6S_2 + 4S_3 - S_4$$

$$S_1 = \frac{7!}{(2!)^3} = \frac{7!}{8} = 630$$

$$S_2 = \frac{6!}{(2!)^2} = \frac{6!}{4} = 180$$

$$S_3 = \frac{5!}{2!} = 60$$

$$S_4 = 4! = 24$$

The total number of restricted words is $4 \cdot 630 - 6 \cdot 180 + 4 \cdot 60 - 24 = 1,656$. So the number of acceptable words is $2,520 - 1,656 = 864$.

Exercise 98. How many ways can you arrange the letters *feminine* so that no letter follows itself?

Answer. The unique letters are (e, i, n, m, f) with multiplicities $(2, 2, 2, 1, 1)$ for a total of 8 letters. The number of unrestricted arrangements is

$$\frac{8!}{(2!)^3} = \frac{8!}{8} = 7! = 5,040$$

Define S_i = number of words where i particular letters appear doubled. The number of restricted words is then

$$\binom{3}{1}S_1 - \binom{3}{2}S_2 + \binom{3}{3}S_3 = 3S_1 - 3S_2 + S_3$$

$$S_1 = \frac{7!}{2!2!} = 1,260$$
$$S_2 = \frac{6!}{2!} = 360$$
$$S_3 = 5! = 120$$

The total number of restricted words is $3 \cdot 1,260 - 3 \cdot 360 + 120 = 2,820$. So the number of acceptable words is $5,040 - 2,820 = 2,220$.

Exercise 99. How many ways can you arrange the letters *muhammadan* so 3 identical letters do not appear together?

Answer. Unique letters and multiplicities are (a, m, d, h, n, u) $(3, 3, 1, 1, 1, 1)$ for a total of 10 letters. The number of unrestricted arrangements is

$$\frac{10!}{3!3!} = 100,800$$

Let S_i be the number of words where i particular letters appear triple. The number of restricted words is then

$$\binom{2}{1}S_1 - \binom{2}{2}S_2 = 2S_1 - S_2$$

$$S_1 = \frac{8!}{3!} = 6,720$$
$$S_2 = 6! = 720$$

So $2S_1 - S_2 = 12,720$ is the total number of restricted words. The number of acceptable words is then $100,800 - 12,720 = 88,080$.

Exercise 100. How many ways can you arrange the letters *muhammadan* so that 2 identical letters do not appear together?

Answer. The unique letters and multiplicities are (a, m, d, h, n, u) $(3, 3, 1, 1, 1, 1)$. Start by looking at the arrangements without the letter a. There are $\binom{7}{3,1,1,1,1} = \frac{7!}{3!} = 840$ such arrangements of which $5! = 120$ have 3 m's in a row (treat 3 m's in a row as one symbol). Take 2 of the m's to be a new letter, then there are 6 letters that can be arranged in $6! = 720$ ways of which $5!$ have 3 m's in a row. The number of arrangements with only 2 m's in a row is then $6! - 5! = 600$. The number of arrangements with all m's single is $840 - 120 - 600 = 120$. There are 3 kinds of arrangements with no a's.

1. 120 arrangements with only *mmm*'s

2. 600 arrangements with only *mm*'s

3. 120 arrangements with only *m*'s

Now combine the *a*'s with these arrangements. For the *mmm* arrangements, 2 of the *a*'s must be used to separate the *m*'s. There are 6 possible positions for the remaining *a*. For the *mm* arrangements, one of the *a*'s must be used to separate the 2 *m*'s. There are 7 possible positions for the remaining 2 *a*'s. For the arrangements with only single *m*'s, there are 8 possible positions for the 3 *a*'s. The number of arrangements of all the letters is therefore

$$\binom{6}{1}120 + \binom{7}{2}600 + \binom{8}{3}120$$
$$= 720 + 21 \cdot 600 + 56 \cdot 120$$
$$= 20,040$$

Exercise 101. Given 20 consecutive numbers, how many ways can you select 2 of them so that they sum to an odd number?

Answer. For 2 numbers to sum to an odd number, one of them must be odd and the other even. In any set of 20 consecutive numbers, 10 will be even and 10 will be odd, so there are $10 \cdot 10 = 100$ ways to make the selection.

Exercise 102. Given 30 consecutive numbers, how many ways can you select 3 to get an even sum?

Answer. Let e represent an even number, and o an odd number, then the possible combinations are:

$$e + e + e = e$$
$$e + e + o = o$$
$$e + o + o = e$$
$$o + o + o = o$$

So the multiplicities of (even,odd) numbers must be $(3, 0)$ or $(1, 2)$ to get an even sum. There are 15 even and 15 odd numbers so the number of ways to select is

$$\binom{15}{3}\binom{15}{0} + \binom{15}{1}\binom{15}{2} = 2,030$$

Exercise 103. How many ways can you collect 20 coins composed of pennies, nickels, and dimes?

Answer. Think of this in terms of distributing 20 identical balls into 3 distinct urns. The 3 urns represent pennies, nickels, and dimes, and a particular distribution of the balls corresponds to a particular collection of coins. The number of ways to distribute n balls into k urns is $\binom{n+k-1}{k-1}$. With $n = 20$ and $k = 3$ this is $\binom{22}{2} = 231$.

Exercise 104. A person has 3 coins and he asks you to guess what they are. Given that the coins could be pennies, nickels, dimes, quarters, half dollars, or dollars, how many guesses do you have to make to guarantee a correct answer?

Answer. The number of possible guesses is equal to the number of ways you can collect 3 coins of 6 possible types. Like the previous problem, this is equal to the number of ways to distribute 3 balls into 6 urns or $\binom{8}{5} = 56$. So if you are very unlucky, then you will have to make 56 guesses before the last one is finally correct.

Exercise 105. How many 5 digit decimal numbers are there (no leading zeros allowed)? In how many of them is every digit odd? In how many is every digit even? In how many are there no digits less than 6? In how many are there no digits greater than 3? How many contain all the digits 1,2,3,4,5? How many contain all the digits 0,2,4,6,8?

Answer. There are $9 \cdot 10 \cdot 10 \cdot 10 \cdot 10 = 90,000$ 5 digit numbers. There are $5 \cdot 5 \cdot 5 \cdot 5 \cdot 5 = 3,125$ numbers with all odd digits. There are $4 \cdot 5 \cdot 5 \cdot 5 \cdot 5 = 2,500$ with all even digits. There are 4 digits ≥ 6, so there are $4^5 = 1,024$ numbers. There are 4 digits ≤ 3, but the first digit cannot be 0 so there are $3 \cdot 4^4 = 768$ numbers. There are $5! = 120$ numbers containing all the digits 1,2,3,4,5. Using all the digits 0,2,4,6,8 there are 4 choices for the first digit and the remaining digits can be arranged in 4! ways, so there are $4 \cdot 4! = 96$ numbers.

Exercise 106. 2 dice with faces numbered 0, 1, 3, 7, 15, 31 are thrown. How many different sums are possible?

Answer. This is equal to the number of ways to distribute 2 identical balls into 6 distinct urns

$$\binom{2+5}{5} = \binom{7}{5} = \frac{6 \cdot 7}{2} = 21$$

Exercise 107. 3 dice with faces 1, 4, 13, 40, 121, 364 are thrown. How many different sums are possible?

Answer. This is equal to the number of ways to distribute 3 identical balls into 6 distinct urns

$$\binom{3+5}{5} = \binom{8}{5} = \frac{6 \cdot 7 \cdot 8}{6} = 56$$

Exercise 108. The post office sells 10 kinds of stamps. How many ways can a person buy 12 stamps? How many ways can a person buy 8 stamps? How many ways can a person buy 8 different stamps?

Answer. For buying 12 stamps, it's like putting 12 identical balls into 10 distinct urns, which can be done in

$$\binom{12+9}{9} = \binom{21}{9} = 293,930$$

ways. Similarly, buying 8 stamps can be done in

$$\binom{8+9}{9} = \binom{17}{9} = 24,310$$

ways. A person can buy 8 different stamps in the same number of ways that 8 things can be selected from 10 things, which is

$$\binom{10}{8} = \frac{9 \cdot 10}{2} = 45$$

ways.

Exercise 109. How many ways can you deal 4 cards to each of 13 players so that each gets one card of each suit? How many ways are there if one person gets one card of each suit and the 12 others each get 4 cards of a single suit?

Answer. Assume a standard deck of 52 cards with 4 suits of 13 cards each. The 13 cards of a suit must be distributed one to a player. Each distribution amounts to a permutation of the cards, and there are 13! permutations per suit. There are 4 suits so the number of ways to distribute the cards is $(13!)^4$. For the second part of the question, the single person can be chosen in 13 ways and then receive the 4 cards in 13^4 ways, so giving one person one card from each suit can be done in 13^5 ways. Now there are 4 suits of 12 cards each left to distribute to the remaining 12 players. Each suit can be divided into 3 groups of 4 cards in $\frac{12!}{(4!)^3}$ ways. This includes permutations of the groups, and we only want combinations, so we must divide this by 3! to get $\frac{12!}{(4!)^3 3!}$ ways. So all 4 suits can be divided into 12 groups of 4 cards (all of the same suit) in $\left(\frac{12!}{(4!)^3 3!}\right)^4$ ways. These 12 groups can be distributed to the 12 players in 12! ways. The number of ways to distribute the cards is then

$$13^5 \frac{(12!)^4}{(4!)^{12}2(3!)^4} 12! = \frac{(13!)^5}{(4!)^{12}2(3!)^4} = \frac{(13!)^5}{2^{40}3^{16}}$$
$$= 197,816,120,269,284,346,830,000,000,000$$
$$\approx 1.98 \times 10^{29}$$

Exercise 110. How many ways can you deal 52 cards to 4 players so that each player has 3 cards each of 3 suits and 4 cards of the remaining suit?

Answer. Assigning the players the suit of which they have 4 cards can be done in 4! ways. Each suit can then be distributed in $\frac{13!}{4! \cdot 3! \cdot 3! \cdot 3!}$ ways and all the cards can be distributed in

$$4! \left(\frac{13!}{4!(3!)^3} \right)^4 = 49,965,764,397,515,366,400,000,000$$

$$\approx 4.997 \text{x} 10^{25}$$

ways.

Exercise 111. How many ways can 18 unique things be distributed to 5 people so that 4 get 4 things and 1 gets 2? How many ways can they be distributed if 3 people get 4 things and 2 get 3 things?

Answer. The set of 18 can be divided into 4 sets of 4 and one set of 2 in $\frac{18!}{4! \cdot 4! \cdot 4! \cdot 4! \cdot 2!}$ ways. For each such division, there are $\binom{5}{4} = 5$ ways the sets can be distributed. The total number of ways to distribute the 18 things is then

$$\frac{5 \cdot 18!}{(4!)^4 2!} = \frac{5 \cdot 17!}{2^{12} \cdot 3^2}$$

$$= 48,243,195,000$$

The set of 18 can be divided into 3 sets of 4 and 2 sets of 3 in $\frac{18!}{4! \cdot 4! \cdot 4! \cdot 3! \cdot 3!}$ ways. For each division, there are $\binom{5}{3} = 10$ ways the sets can be distributed. The total number of ways to distribute the 18 things is then

$$\frac{10 \cdot 18!}{(4!)^3(3!)^2} = \frac{5 \cdot 17!}{2^9 \cdot 3^3}$$
$$= 128,648,520,000$$

.

Exercise 112. There are 14 kinds of things with 2 of each kind. How many different selections are possible?

Answer. For each of the 14 kinds you have 3 choices. You can select 0, 1, or 2 of that kind. The number of selections is then $3^{14} = 4,782,969$. Note that this includes the case of selecting none at all.

Exercise 113. With 20 kinds of things and 9 of each kind, how many different selections can you make?

Answer. Analogous to the previous question there are 10 choices for each thing, so the total number of selections is 10^{20}. Again, this includes the case of selecting none at all.

Exercise 114. Bagatelle is a game similar to billiards. There are 9 balls and 9 different holes. Each hole can hold only one ball, and the object is to get the balls in the holes. In a given state of play, some set of holes will be occupied by balls. If there are 8 white balls and one black ball, how many states can the game be in?

Answer. First consider states where the black ball occupies one of the holes. There are 9 ways this can happen and there are 2 possibilities for the remaining 8 holes, each one can either be occupied or not. The number of states is then $9 \cdot 2^8$. Now look at states where the black ball does not occupy a hole. Let k be the number of white balls occupying holes, then k can range from 0 to 8. The number of ways k balls can occupy 9 different holes is $\binom{9}{k}$. The total number of ways the white balls can occupy the holes is then

$$\sum_{k=0}^{8} \binom{9}{k} = 2^9 - 1$$

The total number of states the game can be in is therefore $9 \cdot 2^8 + 2^9 - 1 = 2,815$.

Exercise 115. For the previous question, if we now have 2 black balls and 7 white balls, how many states can the game be in?

Answer. There can be 0, 1, or 2 black balls occupying

holes, and the number of states for each is:

$$0 : \sum_{k=0}^{7} \binom{9}{k} = 2^9 - \binom{9}{8} - \binom{9}{9} = 2^9 - 10$$

$$1 : 9 \sum_{k=0}^{7} \binom{8}{k} = 9(2^8 - 1)$$

$$2 : \binom{9}{2} \sum_{k=0}^{7} \binom{7}{k} = \frac{9 \cdot 8}{2} 2^7 = 9 \cdot 2^9$$

The total number of states is then $2^9 - 10 + 9(2^8 - 1) + 9 \cdot 2^9 = 7,405$.

Exercise 116. For the previous question, if we now have 1 red ball, 1 green, and 7 white balls, how many states can the game be in?

Answer. With no red or green balls, the number of states is

$$\sum_{k=0}^{7} \binom{9}{k} = 2^9 - \binom{9}{8} - \binom{9}{9} = 2^9 - 10$$

With either a red or green ball, the number of states is

$$2 \cdot 9 \cdot \sum_{k=0}^{7} \binom{8}{k} = 18(2^8 - 1)$$

With both a red and a green ball, the number of states is

$$2 \cdot \binom{9}{2} \sum_{k=0}^{7} \binom{7}{k} = 9 \cdot 8 \cdot 2^7$$

So the total number of states is $2^9 - 10 + 18(2^8 - 1) + 9 \cdot 8 \cdot 2^7 = 14,308$.

Exercise 117. How many ways can you give 27 different books to Alice, Bob, and Cathy so that Alice and Cathy together have twice as many books as Bob?

Answer. Let a, b, and c be the number of books given to Alice, Bob, and Cathy respectively, then $a + b + c = 27$ and $a + c = 2b$. Solving these equations for b and $a + c$ gives, $b = 9$ and $a + c = 18$. 9 books have to be given to Bob, and the remaining 18 are split between Alice and Cathy. Bob can get 9 books in $\binom{27}{9}$ ways, and there are 2 choices for each of the remaining 18 books, so the total number of choices is

$$\binom{27}{9} 2^{18} = 1,228,623,052,800$$

.

Exercise 118. For 99 different things, show that the ratio of the number of ways of selecting 70 to the number of ways of selecting 30 is $\frac{3}{7}$.

Answer. The number of ways to select 70 is $\binom{99}{70}$. The number of ways to select 30 is $\binom{99}{30}$. The ratio is

$$\frac{\binom{99}{70}}{\binom{99}{30}} = \frac{99!}{70!29!} \cdot \frac{30!69!}{99!}$$

$$= \frac{30}{70} = \frac{3}{7}$$

.

Exercise 119. In how many ways can 8 pizzas be delivered by 4 delivery guys so that they each deliver at least one pizza?

Answer. This question is equivalent to "How many ways can you partition a set of 8 pizzas into 4 distinct nonempty subsets?". In terms of putting balls into bins, this is the number of ways to put 8 distinct balls into 4 distinct bins such that each bin has at least one ball. The number of ways to do this is $4! \cdot S(8, 4)$ where $S(8, 4)$ is a Stirling number of the second kind defined as

$$S(8, 4) = \frac{1}{4!} \sum_{i=0}^{3} (-1)^i \binom{3}{i} (3 - i)^8$$

$$= \frac{1}{4!} (4^8 - 4 \cdot 3^8 + 6 \cdot 2^8 - 4)$$

So the number of ways the pizzas can be delivered is

$$4^8 - 4 \cdot 3^8 + 6 \cdot 2^8 - 4 = 40,824$$

Exercise 120. How many ways can you select 3 integers from the set $(1, 2, 3, \ldots, 100)$ so that their sum is divisible by 3?

Answer. The sum will be divisible by 3 under 4 conditions:

1. All integers are of the form $x = 3n$, $n = 1, 2, \ldots, 33$

2. All integers are of the form $x = 3n + 1$, $n = 0, 1, 2, \ldots, 33$

3. All integers are of the form $x = 3n + 2$, $n = 0, 1, 2, \ldots, 32$

4. One integer of each of the above forms.

The number of ways to select the integers is then

$$\binom{33}{3} + \binom{34}{3} + \binom{33}{3} + 33 \cdot 34 \cdot 33 = 53,922$$

Exercise 121. The ratio of the number of ways to select x out of $2x + 2$ to the number of ways to select x out of $2x - 2$ is $\frac{99}{7}$. What is x?

Answer. The ratio is

$$\frac{\binom{2x+2}{x}}{\binom{2x-2}{x}} = \frac{(2x+2)!}{x!(x+2)!} \cdot \frac{x!(x-2)!}{(2x-2)!}$$

$$= \frac{(2x+2)(2x+1)2x(2x-1)}{(x+2)(x+1)x(x-1)}$$

$$= \frac{4(2x+1)(2x-1)}{(x+2)(x-1)}$$

$$= \frac{99}{7}$$

This simplifies to

$$13x^2 - 99x + 170 = 0$$

Solving for x gives $x = 5$.

Exercise 122. If $\binom{n}{3} + \binom{n+2}{3} = \frac{n!}{(n-3)!}$ find n.

Answer. The equation can be written as

$$\frac{n(n-1)(n-2)}{6} + \frac{(n+2)(n+1)n}{6} = n(n-1)(n-2)$$

which simplifies to $2n^2 - 9n + 4 = 0$ then solving for n gives $n = 4$.

Exercise 123. If repetitions are allowed, show that the number of ways to select n things out of $m+1$ is the same as the number of ways to select m things out of $n+1$.

Answer. Another way of saying this is that the number of ways to put n identical balls into $m + 1$ distinct urns is the same as the number of ways to put m identical balls into $n + 1$ distinct urns. In both cases the number is

$$\binom{n+m}{n} = \binom{n+m}{m}$$

.

Exercise 124. If you throw n dice how many different results are possible?

Answer. This is equivalent to asking how many ways n identical balls can be put into 6 distinct urns. The number is $\binom{n+5}{5}$.

Further Reading

- *A Course in Enumeration*, Martin Aigner

- *102 Combinatorial Problems*, Titu Andreescu

- *Schaum's Outline of Theory and Problems of Combinatorics*, V.K. Balakrishnan

- *Proofs that Really Count: The Art of Combinatorial Proof*, Benjamin and Quinn

- *A Combinatorial Miscellany*, Bjorner and Stanley

- *Combinatorics of Permutations*, Miklos Bona

- *A Walk Through Combinatorics: An Introduction to Enumeration and Graph Theory*, Miklos Bona

- *Advanced Combinatorics*, Louis Comtet

- *Analytic Combinatorics*, Flajolet and Sedgewick

- *Combinatorial Problems and Exercises*, Laszlo Lovasz

- *Discrete Mathematics*, Laszlo Lovasz

- *Combinatorics*, Russell Merris

- *Combinatorial Identities*, John Riordan

- *Enumerative Combinatorics, Vol 1, 2nd ed*, Richard P. Stanley

- *Choice and Chance with 1000 Exercises, 5th edition*, William Allen Whitworth

- *Generatingfunctionology*, Herbert S. Wilf

- *Challenging Mathematical Problems With Elementary Solutions, Vol. 1*, A.M. Yaglom and I.M. Yaglom

Acknowledgements

We'd like to thank our parents for everything that we have.

We thank the makers and maintainers of all the software we've used in the production of this book: the Emacs text editor, Emacs Calc, the LaTex typsetting system, Inkscape, Evince document viewer, Maxima computer algebra system, gcc, Guile, awk, z-shell, and the Linux operating system.

Stefan Hollos and **J. Richard Hollos** are physicists by training, and enjoy anything related to probability. They are the authors of

- **Probability Problems and Solutions**

- **The Coin Toss: Probabilities and Patterns**

- **Pairs Trading: A Bayesian Example**

- **Simple Trading Strategies That Work**

- **Bet Smart: The Kelly System for Gambling and Investing**

- **The QuantWolf Guide to Calculating Bond Default Probabilities**

- **The Mathematics of Lotteries: How to Calculate the Odds**

- **Signals from the Subatomic World: How to Build a Proton Precession Magnetometer**

They are brothers and business partners at Exstrom Laboratories LLC in Longmont, Colorado. The websites for their work are Exstrom.com and QuantWolf.com.

Thank you for buying this book.

If you bought this book from a retailer, register it with us and receive news on updates, special offers, and related products. Just go to

http://www.abrazol.com/books/combinatorics1/

and enter your email address.